ANALYSIS *of* GEOLOGIC STRUCTURES

ANALYSIS *of*

NEW YORK

John M. Dennison

University of North Carolina at Chapel Hill

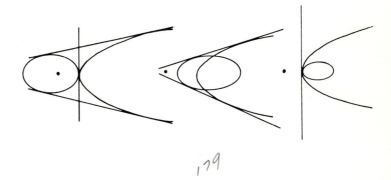

179

GEOLOGIC STRUCTURES

W · W · NORTON & COMPANY · INC ·

Contents

Preface

THIS MATERIAL IS PREPARED CHIEFLY as a laboratory workbook for undergraduate students in structural geology. It also is designed for use later in connection with geologic field work and to serve as an introduction to some specialized techniques for graduate and professional levels of work. Since the topical coverage is too broad for a single course, the fundamental ideas are in the first seven chapters, followed by advanced topics that can be used selectively.

The book is an outgrowth of a pleasant friendship with a former teacher at West Virginia University, John C. Ludlum. In 1957, we prepared *Structural Geology Laboratory Manual* (Ludlum and Dennison, Edwards Bros., Ann Arbor, Michigan). A revised edition appeared in 1960. Administrative duties prevented Dr. Ludlum from further participation, so the present book is my rewriting and expansion of many of our original ideas, plus much new material. I wish to assume total responsibility for any deficiencies that occur in the present book.

Structural geology is taught in many ways and at different stages in college and university curricula. The approach here is to concentrate on developing the student's reasoning powers concerning the geometry of rocks in nature, while using a minimum of descriptive terminology. This material is prepared for use in special laboratory or for assignment in conjunction with a structural geology lecture course that immediately follows historical geology in the curriculum. In this sequence it will serve as a transition between the comfortable generalizations and facts of physical and historical geology on one hand and the more difficult field relations of rocks in four dimensions on the other. If a student has been trained to visualize rock sections in their spatial and time dimensions prior to an intensive study of field geology, frustrations will be minimized and chances are increased for his developing a lifelong enjoyment of field work. Proper placement of rocks in their field relations is still the greatest frontier for the advancement of geologic knowledge.

The trigonometric approach to geologic geometry is emphasized more here than in many texts. Descriptive geometry solutions to problems are easier to visualize, but practical application in field work shows that the trigonometric method is actually simpler. Trigonometric solutions are quicker, and an accuracy of three significant figures (all that the quality of the original data warrants) can generally be obtained in the field with the aid of a slide rule. The geologist is then able to make a visual comparison of his calculations with the field situation, thereby checking the reasonableness of the solution. It is essential in all field work to obtain sufficient data or samples to solve the problem at hand before returning to the office or laboratory.

Derivations of formulae are in small print, because these are for reference, and generally need not be memorized in detail, but only understood. Procedures are outlined for solving particular types of problems, and in the exercises answers are included for the first of most new categories of problems to serve as a check for the solution.

Discussions with several geologists have contributed significantly to this material. These include William Dudley, Kenneth Hasson, Michael L. Jones, Joseph Mengel, Edward Norwood, and the late Stanley A. Tyler. I appreciate their constructive comments.

JOHN M. DENNISON

Chapel Hill, North Carolina

Recommended Supplies

[*Items marked with asterisk should be available for every exercise.*]

*A number 12 (12 inches on long side) transparent triangle with 30, 60, and 90-degree angles.

*A transparent protractor of 6-inch base (diameter) with ½-degree divisions.

*A triangular, plastic-faced engineer's scale having inches divided into units that are multiples of 10.

*A 3H or 4H drawing pencil and suitable eraser. A 2H or F pencil is better for calculations and general writing.

*A roll of drafting tape of the crinkled tan or gray type (not the transparent cellophane or plastic type). The smallest-size roll will be more than enough for use in all problems.

A long T-square, 35 to 41 inches in length, for scale enlargement and point projection in construction of structure sections.

A pair of steel dividers about 6 inches long. May be part of a drafting set.

A slide rule, either straight or circular. A circular slide rule is very convenient for field use. The data used for all of these types of structural geology problems are sufficiently imprecise that slide-rule accuracy is adequate.

A ruling pen of the sort in ordinary drafting sets, for use in drawing the base and end lines of the large structure section. A Rapidograph pen is a little more expensive, but far more versatile than a ruling pen.

A compass for drawing circles in pencil.

A straight pen for fine-line work on structure sections. C. H. Hunt Pen Co. of Camden, N. J., makes a satisfacory set consisting of a penholder and one dozen crowquill pens. A 00-size Rapidograph pen is equally satisfactory, but more expensive.

A set of colored pencils for coloring different formations of structure sections. The Mongol #741 assortment of 12 pencils by the Eberhard Faber Pencil Co., Brooklyn, N. Y., is satisfactory.

One dozen artist's blending stumps in pencil-shaped form of rolled paper for use in smoothing out the colored pencil work on structure sections.

A bottle of black drawing ink (India-ink type) for finishing the large structure section exercise.

Paper of several kinds:

$8\frac{1}{2} \times 11$-inch cross-section paper for preliminary structure sections.

12×36-inch sheet of cross-section paper for large, finished structure section problem.

$8\frac{1}{2} \times 11$-inch rough, canary yellow sheets for general computation and problem work.

$8\frac{1}{2} \times 11$-inch onionskin or other transparent sheets for overlay work in stereonet solutions and fault problems.

ANALYSIS *of* GEOLOGIC STRUCTURES

CHAPTER 1

True and Apparent Dips

THE *true dip* OF A BED is the direction and amount of inclination of the bedding plane with respect to the horizontal, measured in a vertical plane at right angles to the strike (Fig. 1–1). It can also be defined as the direction and inclination of the steepest line that can be drawn on the bedding plane. Strike is the compass direction of a level line on the bedding plane, conventionally expressed as degrees east or west of north (rather than using the southern half of the compass circle).

FIG. 1–1   Strike and dip of strata.

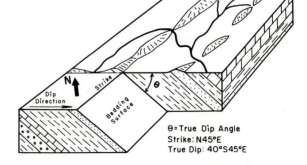

θ = True Dip Angle
Strike: N45°E
True Dip: 40°S45°E

An example of the conventional method for recording dip is 42°N63°E (meaning that the bed is inclined 42° from the horizontal in a direction N63°E). The strike of this bed would be recorded as N27°W. Given the dip direction, the strike direction is readily calculated by adding 90° to or subtracting 90° from the dip direction to give the trend of the strike line in the northern semicircle. Occasionally, bedding attitude is expressed as N27°W 42°NE, indicating true strike direction, dip amount, and the quadrant of dip direction. This last form is most frequently used to record field measurements, where strike direction and dip amount are measured directly, and the dip direction is obtained later by calculation.

The apparent angle of dip in a vertical cross section drawn not perpendicular to the strike is less than the true dip angle measured in a cross section perpendicular to the strike (Fig. 1–2).

Strike direction and true and apparent dips can be determined for any rock mass that has an approximate plane surface such as a bedding surface, joint, fault surface, border of a tabular igneous intrusion, an unconformity, cross-bedding, schistosity in metamorphic rocks, or flow layers in igneous rocks. In the following explanations such a plane surface will be consistently called a bed for convenience. Of course, the methods of analysis apply to these other plane surfaces as well as beds.

There are many situations in which a geologist may know the true dip of a bed, but wishes to know the apparent dip of the bed on the sides of a drift or in a quarry wall that trends in a direction other than that of the true dip. On the other hand, the geologist may find it impossible to take a true strike and dip reading on bedding exposed in vertical quarry walls or in a mine, but the true dip can be easily established if he can sight the apparent inclinations of the beds in the direction of trends of the walls. Graphical, trigonometric, stereonet, or alignment-diagram methods are most commonly used to solve these kinds of problems. Personal preference of the individual determines the method used.

Graphical Methods

Figure 1–2 illustrates a true dip (θ) and two apparent dips (α and β) in a three-dimensional diagram. Plane AJCE is horizontal, and plane AFGH is the bedding plane. Line FGH is the strike, and line JCE parallels it. Angle JAF (angle α) is the apparent dip measured in the direction of line AJ, and angle EAH (angle β) is the apparent dip measured in the direction of line AE. Angle CAG is the true dip amount (angle θ) in the true dip direction (AC).

True Dip from Two Apparent Dips / If two apparent dips are known, such as those shown in three dimensions in Figure 1–2, the attitude of the true dip can be determined. The graphical solution is illustrated in Figure 1–3, where the following problem is solved. *Given two apparent dips, one with an apparent dip of 14° in a direction N83°W, and the other an apparent dip of 15° in a direction S36°W. What is the true dip direction and amount?*

First construct east-west and north-south coordinates to aid in plotting directions. Lines AJ and AE are the plots of the two apparent dip directions, and angles JAF and EAH are the respective apparent dip angles. Planes JAF and EAH have been rotated 90° clockwise and 90° counterclockwise respectively, as viewed from J and E, from the positions as shown in Figure 1–2. Because lines JCE and FGH of Figure 1–2 are

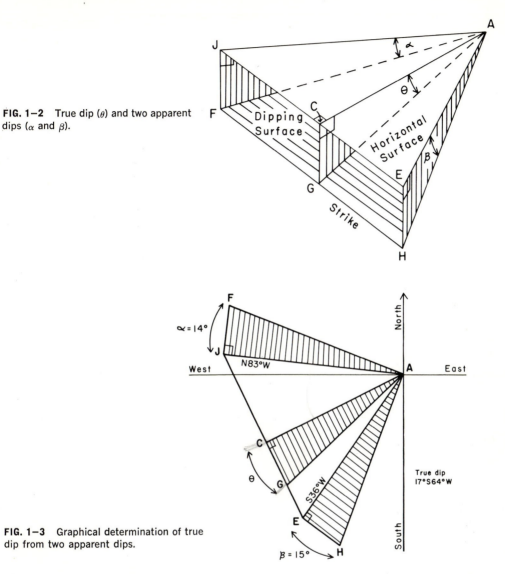

FIG. 1–2 True dip (θ) and two apparent dips (α and β).

FIG. 1–3 Graphical determination of true dip from two apparent dips.

parallel and lie in a vertical plane, the vertical distance of the strike line (FGH) from the parallel line (JCE) in the horizontal plane is constant, and therefore JF, CG, and EH should be of equal length as plotted in Figure 1–3. Lines JF and EH are drawn at right angles to the apparent dip directions, with points J and E located so that lines JF and EH are the same length, fixed at some arbitrary value for convenience. Points J and E, thus determined, establish the direction of the true strike of the bed. AC is drawn perpendicular to strike line JE, so that the true dip direction is given by AC (S64°W). Lay off CG equal in length to EH. The true dip amount ($\theta = 17°$) is angle CAG.

The true dip direction in Figure 1–3 lies between the two apparent dip directions. This is not always true, however. The method of solution is essentially the same, nevertheless. Consider this example. *Given one apparent dip of 16° in a direction N85°W and another apparent dip of 30° in a direction of S70°W.* The graphical solution for true dip is illustrated in Figure 1–4, resulting in a true dip of 38° in a direction S28°W.

FIG. 1–4 Graphical determination of true dip when it does not lie between two apparent dips.

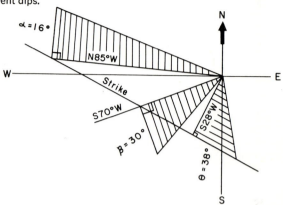

Apparent Dip from True Dip / The apparent dip in any direction can be obtained if the true dip is known. *Consider an example in which the true dip is 17° S65°W, and it is desired to find the apparent dip in a direction N70°W.* In Figure 1–5 line AC is the true dip direction, and CAG is the true dip angle. Draw a strike line (TG) perpendicular to the true dip. The angle between the true and apparent dip direction (CAT) is laid off. Point T is determined by the intersection of the apparent dip direction (line AT) and the strike (GT). TU, which equals CG, is laid off perpendicular to AT at point T. Angle TAU is the apparent dip angle in the direction of AT, a value of 12° in this example.

FIG. 1–5 Determination of apparent dip from true dip.

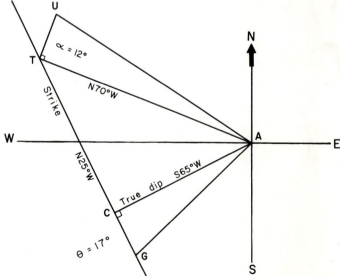

It is difficult to study very gentle dips by the ordinary graphical method described in the preceding paragraphs. Dip angles are actually an expression of the vertical change of position of a bed in a certain horizontal distance. If the dips are expressed and laid off as vertical displacement per horizontal distance instead of angular measurement, the vertical change of position can be exaggerated by some constant factor throughout an entire problem without changing the results of the solution. For example, if the true dip is 30 feet per mile, it can be multiplied by a factor of 100 and can be used in a drawing as 3,000 feet per mile. If an apparent dip from this comes out as 2,100 feet per mile, this value can be reconverted to yield the actual apparent dip of 21 feet per mile.

Exercises /

1. The true dip of a sandstone is 27°N30°E. What will be the apparent dip along a proposed railroad cut trending N72°E? (Answer: 20°N72°E)
2. A coal seam has a true dip of 10°S37°W. What will be the slope of a drift trending due west in this seam?
3. The true dip of relic bedding in a metamorphosed sediment is 35°S42°W. What will be its apparent dip on vertical slaty cleavage striking N80°W?
4. The true dip of a limestone bed is 12°N48°W. It is desired to mine the limestone by digging a horizontal drift trending straight into the mountainside in a direction N13°E. What would be the inclination of the bedding along the sides of the drift?
5. It is planned to make a highway cut through rock with closely spaced jointing. The attitude of the joints is N60°W 75°N30°E. The proposed cut would have essentially vertical walls and would trend N75°W. What would be the apparent dip of the trace of the joints on the vertical surface of the cut? On which side of the highway would there be great danger of slides, and why?
6. The apparent dip of limestone strata in a quarry wall is 17°N69°E. In another wall of the quarry, the apparent dip is 13°N20°W. What is the true dip of the limestone? (Answer: 32½°N49½°E, rounded to 32°N49°E)
7. The apparent dip of the contact between a sill and the country rock is 14°S3°W. Another apparent dip of the same contact is 37°S45°W. What is the true dip of the contact?
8. A vein exposed along a vertical creek bank has an apparent dip of 55°N37°W. In another exposure along a vertical quarry wall trending S60°W the apparent dip is 82°S60°W. What is the true dip and strike of the vein?
9. The apparent dip of a lamprophyre dike cutting across a granite quarry wall is 39°S65°W. The apparent dip across another wall of the quarry is 52°S10°E. What is the true dip of the dike?
10. Uniform cross-bedding is exposed on two sides of a block bounded by joints. On one joint surface the apparent dip of the cross-bedding is 5°N16°E. It is 12°N88°E on the other joint surface. What is the true attitude of the cross-bedding at that point?

11. A fault crops out at three points. Point A is 2,276 feet above sea level. Point B (1,652 feet altitude) is 1,500 feet N16°W of point A. Point C (2,510 feet altitude) is 1,700 feet S39°W of point A. The distance between the two points is horizontal distance. What is the true dip of the fault surface, assuming that it is a plane?

12. A fault crops out at three known points. Point A is 1,742 feet above sea level. Point B (1,707 feet altitude) is 1,200 feet N77°W of point A. Point C (1,800 feet altitude) is 900 feet N23°W of point A. The distance between points is horizontal distance. What is the true dip of the fault surface, assuming that it is a plane?

13. Three vertical holes were drilled to the top of a copper-bearing lava flow. Hole A has a surface altitude of 1,116 feet and a depth of 192 feet. Hole B (surface altitude 1,165 feet, depth 641 feet) is located 432 feet N35°W of hole A. Hole C (surface altitude 1,129 feet, depth 210 feet) is located 355 feet N50°E of hole A. What is the attitude of the flow, assuming that the top surface is a plane?

14. Three vertical cores are drilled to the top of a limestone containing galena and sphalerite. Hole A has a surface altitude of 963 feet and a depth of 192 feet. Hole B (surface altitude 987 feet, depth 55 feet) is located 432 feet N27°W of hole A. Hole C (surface altitude 950 feet, depth 250 feet) is located 380 feet N34°E of well A. Calculate the attitude of the top of the ore-bearing stratum, assuming that the limestone bedding surface is a plane.

15. A true regional dip reading of 17°N36°W is obtained for the top of a fire-clay horizon. A vertical shaft is to be sunk to the clay at a point where it is 70 feet below the surface, and the clay is to be mined in drifts radiating from that point. If the ground surface is level, at what distance due north of the exposure should the vertical shaft be excavated? (Answer: 283 feet)

16. A true regional dip of 12°S14°W is obtained for a limestone stratum bearing galena and sphalerite. A vertical shaft is to be sunk to the ore-bearing stratum at a point where it is 200 feet below the surface. It is desired to locate the shaft along a railroad trending S67°W from the exposure of the stratum. At what distance along the railroad from the surface exposure should the shaft be located? Assume that the ground surface is level.

17. The true dip of a coal seam measured at an outcrop is 12°S79°W. It is desired to locate a shaft mine exactly one-half mile N30°W of the exposure. If the ground surface is level, how deep will the vertical shaft have to be in order to reach the coal?

18. A dip-slope hill is composed of resistant sandstone with a dip of 13°S42°E. The crest of the hill is 420 feet above the base. A proposed railroad will extend in a northeasterly direction with a slope of 2°. What will be the exact direction of the railroad? How long will the grade be?

19. The top of the Tuscaloosa Formation has a true dip of 120 feet per mile in a direction S45°W. What is its apparent dip in a direction N75°W?

20. The Yorktown Formation dips 45 feet per mile in a due east direction. What is its apparent dip in a direction N53°E?

21. The Mahantango Formation is a clastic wedge of sediments with isopachs trending

north-south. It thins from east to west from 720 feet to 500 feet in a distance of 8.0 miles. What would be the apparent rate of thinning along a strike belt trending N25°E? Express the apparent rate of thinning as feet per mile. Would the formation appear to thin to the northeast or to the southwest?

Trigonometric Methods

The following derivation can be obtained from Figure 1–2 to show the relationship between directions and amounts of true and apparent dips.

$JF = CG$.

$$\text{Tan } JAF = \frac{JF}{AJ} = \frac{CG}{AJ}.$$

$$\text{Tan } JAF = \frac{CG}{AC \ (\text{Sec } JAC)}.$$

$$\text{Tan } JAF = \frac{AC \ (\text{Tan } CAG)}{AC \ (\text{Sec } JAC)} = \frac{\text{Tan } CAG}{\text{Sec } JAC}.$$

$\text{Tan } JAF = (\text{Tan } CAG) (\text{Cos } JAC)$.

Tan apparent dip angle = (Tan true dip angle) (Cos angle between true and apparent dip directions). (1)

Another form of this equation is

Tan true dip angle = $\dfrac{\textbf{Tan apparent dip angle}}{\textbf{Cos angle between true and apparent dip directions}}.$ (2)

It should be evident that Formulae 1 and 2 can be used only:
1. when the directions of true and apparent dip are known, the amount of one dip is also known, and the amount of the other dip is desired, or
2. when the true and apparent dip amounts are known, the direction of either the true or apparent dip is known, and the other dip direction is desired.

True Dip from Two Apparent Dips / The following derivation based on Figure 1–2 yields a formula for determining the true dip direction from two known apparent dip directions and amounts.

$JF = CG = EH = m$.

$\dfrac{AE}{EH} = \text{Cot } EAH$.

$AE = EH (\text{Cot } EAH) = m (\text{Cot } EAH)$.

$AC = AJ (\text{Cos } JAC)$.

$AC = AE (\text{Cos } CAE)$.

Angle $CAE =$ Angle $JAE -$ Angle JAC.

$AC = AE \text{ Cos } (JAE - JAC)$.

$\therefore AJ (\text{Cos } JAC) = AE \text{ Cos} (JAE - JAC)$.

$m (\text{Cot } JAF) (\text{Cos } JAC) = m (\text{Cot } EAH) [\text{Cos } (JAE - JAC)]$.

But

Cos $(JAE - JAC) = (Cos\ JAE)(Cos\ JAC) + (Sin\ JAE)(Sin\ JAC)$.

$\therefore (Cot\ JAF)(Cos\ JAC) = (Cot\ EAH)[(Cos\ JAE)(Cos\ JAC) + (Sin\ JAE)(Sin\ JAC)]$.

Divide both sides of equation by (Cot EAH).

$\dfrac{(Cot\ JAF)(Cos\ JAC)}{(Cot\ EAH)} = (Cos\ JAE)(Cos\ JAC) + (Sin\ JAE)(Sin\ JAC)$.

$(Cot\ JAF)(Cos\ JAC)(Tan\ EAH) = (Cos\ JAE)(Cos\ JAC) + (Sin\ JAE)(Sin\ JAC)$.

Transposing as follows:

$(Sin\ JAE)(Sin\ JAC) = (Cot\ JAF)(Cos\ JAC)(Tan\ EAH) - (Cos\ JAE)(Cos\ JAC)$.

Divide both sides of equation by (Sin JAE) (Cos JAC).

$\dfrac{(Sin\ JAE)(Sin\ JAC)}{(Sin\ JAE)(Cos\ JAC)} = \dfrac{(Cot\ JAF)(Cos\ JAC)(Tan\ EAH)}{(Sin\ JAE)(Cos\ JAC)} - \dfrac{(Cos\ JAE)(Cos\ JAC)}{(Sin\ JAE)(Cos\ JAC)}$.

$\dfrac{Sin\ JAC}{Cos\ JAC} = \dfrac{(Cot\ JAF)(Tan\ EAH)}{Sin\ JAE} - \dfrac{Cos\ JAE}{Sin\ JAE}$.

$Tan\ JAC = \dfrac{(Cot\ JAF)(Tan\ EAH) - (Cos\ JAE)}{Sin\ JAE}$.

$Tan\ JAC = (Csc\ JAE)[(Cot\ JAF)(Tan\ EAH) - (Cos\ JAE)]$.

(**Tan angle between first apparent dip and true dip directions**) = (**Csc angle between two apparent dip directions**) [(**Cot first apparent dip angle**)(**Tan second apparent dip angle**) − (**Cos angle between two apparent dip directions**)]. (3)

Formula 3 is used to determine the angle between the true dip and one of two known apparent dips, thus permitting computation of the true dip direction. The true dip amount can be calculated by using Formula 2. If the computed angle between one apparent dip direction and the true dip direction is larger than the angle between the two apparent dip directions, the true dip does not lie between the two apparent dip directions. In such a case the true dip direction is in the general direction of and beyond the direction line for the steeper of the two apparent dips, with the calculated angle being the directional value between the true dip and the more gentle of the two apparent dips.

Tangent Vector Method / A simple combined trigonometric and graphic solution of the true and apparent dip problems is the tangent vector method (Hubbert, 1931). The procedure uses lines drawn in the true and apparent dip directions with lengths proportional to the tangents of the respective dip angles. Application of the method is illustrated with the problems solved in another manner in Figures 1–3 and 1–5.

Given a true dip of 17°S65°W, what is the apparent dip in a direction N70°W? Figure 1–6 illustrates the tangent vector solution. Lay off a N-S and E-W axis, and then plot a line through the intersection extending S65°W. Make the length (AB) of that line proportional to the tangent of 17° (AB = .306 units). Draw a second line through point A extending in a direction N70°W. Pass a line from the end of the true dip vector (point B) in a direction perpendicular to the apparent dip direction, so that angle BCA is a right angle. The length of line CA is proportional to the tangent of the

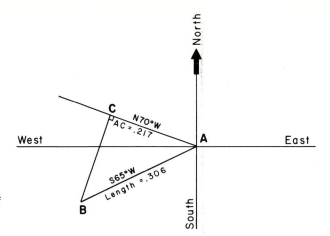

FIG. 1–6 Tangent vector solution of apparent dip from true dip.

apparent dip angle (CA = .217). Converting this to degrees, the apparent dip is 12°. This agrees with the values obtained in Figure 1–5 and calculated by Formula 1.

In the example of *solving for true dip from two apparent dips of 14°N83°W and 15°S36°W*, the tangent vector solution is shown in Figure 1–7. Lay off N-S and E-W coordinates, and then plot a line extending N83°W from the coordinate center with a length proportional to tan 14° (AB = .249 units). A line extending S36°W from the coordinate center is proportional to tan 15° (AC = .268 units). From point B draw a line perpendicular to AB, and draw a line perpendicular to AC at point C. The intersection of these last two lines at G determines the direction of the true dip along line AG. The length of AG (AG = .299 units) is the tangent of the true dip angle, which is 16°44′. This compares favorably with a true dip value of 17°S64°W in Figure 1–3 and with a trigonometric determination of 16°48′S63°13′W resulting from Formulae 3 and 2.

Tangent vectors are easier to use than Formulae 2 and 3 for obtaining true dip from two apparent dips and are probably more accurate than the purely graphical solution of such a problem, because fewer lines must be laid off with specific angular relationships. The tangent vector method is really no easier to use than the strictly trigonometric procedure for calculating apparent dip from a given true dip.

FIG. 1–7 Tangent vector method of determining true dip from two apparent dips.

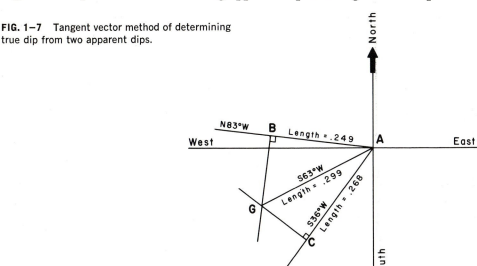

Nettleton (1931) suggested that a plot of dip directions with length of dip line related to dip amount could be used to get a best average dip from several dip readings on a bed and could help one spot erroneous readings. Nevin (1949, pp. 344–345) illustrates the procedure, corrected to use the tangent vector method.

Alignment Diagram Method

Various tables and graphical devices have been developed to facilitate the solution of true and apparent dip problems. The most commonly encountered situation is that of finding apparent dip in a desired direction, with the true dip already known.

Tabulated values of apparent dips in directions at various angles from the true dip are presented by Lahee (1961, p. 857).

Even simpler to use is a graphical device called a nomogram. Figure 1–8 is a nomogram that is especially helpful when a structure section is drawn at an oblique angle to the trends of the formation and it is necessary to convert true dips to apparent dips in the plane of the structure section. In the example shown by the dashed line in Figure 1–8 the formations of the area have a true dip of 43°, and the apparent dip in a direction 35° from the strike is desired. The apparent dip is read from the middle scale as 28°. Notice that for this alignment diagram, or nomogram, the apparent dip direction is measured from the strike.

The nomogram in Figure 1–8 is not very accurate for gentle dips, those with less than 5° true dip amount. Fisher (1937, p. 342) has prepared a chart showing apparent dips when the true dip angle is 5° or less. His chart can be used with dip expressed in degrees or in feet per mile.

Probably the most versatile diagrammatic method is the slide rule developed by L. R. Satin (1960). His apparent dip computer is shown in Figure 1–9. Cut out this computer on the outer circumference, and cut with a razor blade along the circular line between the true and apparent dip scales. Mount the outer ring on a stiff backing, and attach the inner dial so that it can rotate about the center (marked position). This device is simple to use and compact. It is more accurate than the alignment diagram in Figure 1–8.

Stereonet Method

Another device that has been used to determine apparent dip when the true dip is known is the stereonet. This geometric projection of angles in a three-dimensional space solves many types of angular relation geologic problems, and Chapter 11 is devoted to stereonet solutions. Stereonets can also be used to determine the true dip direction and amount, given two apparent dips.

FIG. 1—8 Nomogram or alignment diagram for use in true and apparent dip problems.(*Palmer 1919.*)

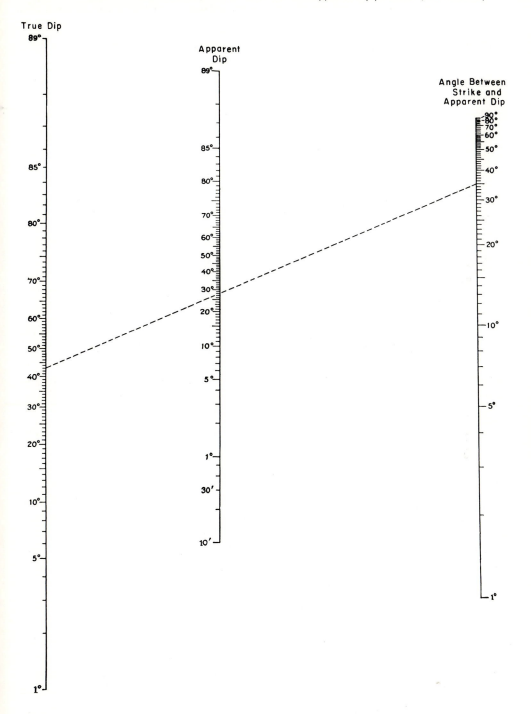

Exercises / These problems should be solved by the graphical method, checked trigonometrically, and verified again by using the alignment diagram or apparent dip computer.

22. Strata viewed along a quarry face trending N36°E appear to be horizontal. Along another face the apparent dip is 23°N30°W. Determine the true dip.

23. Limestone exposed along a quarry wall trending N27°W has a horizontal attitude. Along another face the dip appears to be 42°S18°W. What is the true dip?

24. The attitude of a formation contact line is N89°W 32°NE as determined from a geologic map. What would be the dip of the formation contact in a structure section trending N45°E?

25. The attitude of the Lower Kittanning coal is N10°E 21°S80°E. What would be the dip of this coal in a structure section trending S25°E?

26. Two apparent dips are taken on a sandy shale. One is 10°N60°W, and the other is 25°N30°W. Find the true dip.

27. Two apparent dips on a sill are 15°S27°E and 5°S5°W. What is the true dip of the sill?

28. The apparent dip of a fault surface in a drift is 25°S65°W. The apparent dip in a branch drift is 18°S20°E. What is the true dip of the fault surface, assuming that it is a plane?

29. The apparent dip of a dike in a quarry is 32°S45°W. The apparent dip of the dike on another wall of the quarry is 17°N58°W. What is the true dip of the dike, assuming that it is planar?

30. Three outcrops of the superface of the Bedford Limestone are located. Outcrop A is 877 feet in altitude. Outcrop B (altitude 843 feet) is located 3,250 feet map distance N23°E of outcrop A. Outcrop C (altitude 830 feet) is 3,940 feet map distance N47°W of point A. Determine the true dip of the limestone, using apparent dips.

31. Three outcrops of the subface of the Sundance Formation are located by plane table mapping. Outcrop A is 6,427 feet altitude. Outcrop B (altitude 7,115 feet) is 7,450 feet N87°W of outcrop A. Outcrop C (altitude 6,745 feet) is 2,980 feet S45°W of point A. Determine the true dip of the formation by using apparent dips.

32. Three exposures of a fire clay are marked by springs along the sides of a valley. Point A is 1,876 feet above sea level. Point B (elevation 2,310 feet) is located 1,850 feet N17°E from point A. Point C (elevation 1,999 feet) is located 2,600 feet S85°E from point A. What is the true dip of the fire clay? Use apparent dips.

33. A coal seam is exposed at three points along the sides of a valley. Point A is 1,450 feet above sea level. Point B (altitude 1,430 feet) is 4,250 feet S42°W of point A. Point C (altitude 1,395 feet) is 3,120 feet S15°W of point A. What is the true dip of the coal seam?

34. Vertical cores are drilled to the top of a sill at three points on the map of Figure 1–10. The top of the sill is reached at a depth of 117 feet in hole A, at a depth of 872 feet in hole B, and at a depth of 560 feet in hole C. Use apparent dips to find the attitude of the sill.

FIG. 1–9 Apparent-dip computer.
(From "Apparent-dip Computer" by L. R. Satin, *Bulletin of the Geological Society of America,* 1960.)

Cut out computer on the circular line between the true dip and apparent dip scales. Mount the outer ring on a stiff backing and attach the inner dial at the center marked position. Instructions for use are printed on the dial face.

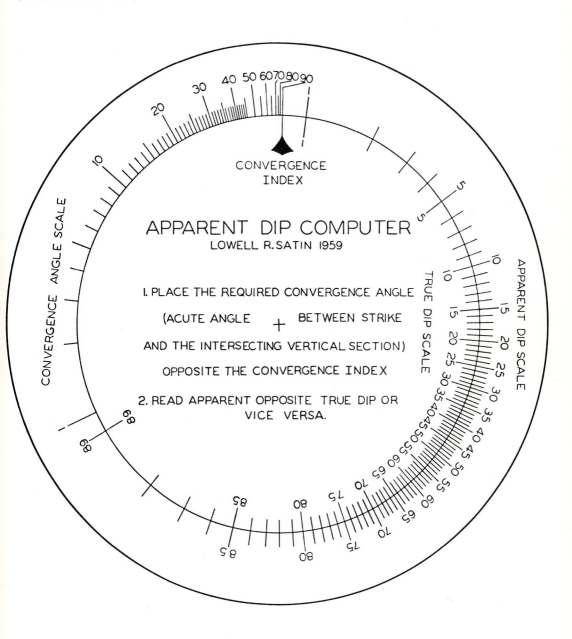

35. The top of the Selma Chalk is penetrated by the three wells shown on the map of Figure 1–10. The depth to the top of the formation is 1,525 feet in well A, 1,555 feet in well B, and 1,617 feet in well C. What is the strike and dip of the chalk? Use apparent dips.

36. Core holes are drilled through the Clinton hematite iron ore at three points on the map of Figure 1–10. The base of the ore bed is reached at a depth of 680 feet in well A, 1,213 feet in well B, and 1,148 feet in well C. What is the attitude of the hematite bed? Use apparent dips.

37. The top of the Midway Formation is penetrated by the three wells shown on the map of Figure 1–10. The depth to the top of the formation is 2,320 feet in well A, 2,415 feet in well B, and 2,363 feet in well C. Find the strike and dip of the Midway Formation, using apparent dips.

FIG. 1–10   Map for use with Exercises 34–37.

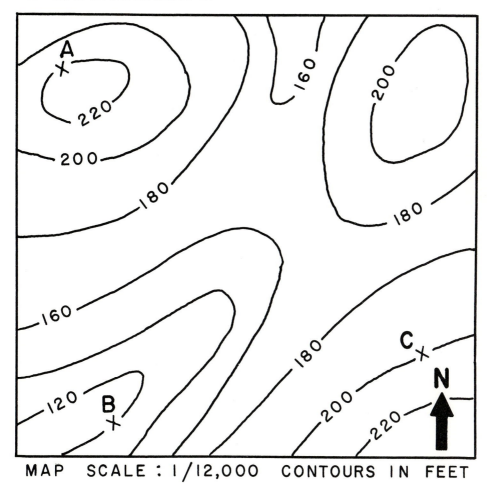

MAP   SCALE : 1/12,000   CONTOURS IN FEET

References

Fisher, D. J., 1937, Some dip problems: Am. Assoc. Petroleum Geologists Bull., v. 21, p. 340–351.

Herold, S. C., 1933, Projection of dip angle on profile section: Am. Assoc. Petroleum Geologists Bull., v. 17, p. 740–742.

Hewett, D. F., 1912, A curved network for true and apparent dips: Econ. Geology, v. 7, p. 190–191.

Hubbert, M. K., 1931, Graphic solution of strike and dip from two angular components: Am. Assoc. Petroleum Geologists Bull., v. 15, p. 283–286.

Hughes, R. J., 1960, A derivation of Earle's formula for the calculation of true dip: Southeastern Geology, v. 2, p. 43–48.

Johnson, C. H., 1936, Nomographic solution for apparent dip in vertical section not perpendicular to strike: Am. Assoc. Petroleum Geologists Bull., v. 20, p. 816–818.

Johnson, L. H., 1952, Nomography and empirical equations: New York, John Wiley & Sons, 150 p.

Kitson, H. W., 1929, Graphic solution of strike and dip from two angular components: Am. Assoc. Petroleum Geologists Bull., v. 13, p. 1211–1213.

Lahee, F. H., 1919, Graphic determination of dip components where dips are measured in feet per mile: Econ. Geology, v. 14, p. 262–263.

———, 1961, Field geology: New York, McGraw-Hill Book Co., 6th ed., 926 p.

Levens, A. S., 1959, Nomography: New York, John Wiley & Sons, 2nd ed., 296 p.

Nettleton, L. L., 1931, Graphic solution of strike and dip from two angular components: Am. Assoc. Petroleum Geologists Bull., v. 15, p. 79–82.

Nevin, C. M., 1949, Principles of structural geology: New York, John Wiley & Sons, 4th ed., 410 p.

Palmer, H. S., 1919, New graphic method for determining the depth and thickness of strata and the projection of dip: U.S. Geol. Survey, Prof. Paper 120, p. 123–128.

Rich, J. L., 1932, Simple graphical method for determining true dip from two components and for constructing maps from dip observations: Am. Assoc. Petroleum Geologists Bull., v. 16, p. 92–95.

Satin, L. R., 1960, Apparent-dip computer: Geol. Soc. America Bull., v. 71, p. 231–234.

Threet, R. L., 1957, Automatic dip-component computer for use with Brunton compass: Am. Assoc. Petroleum Geologists Bull., v. 41, p. 2752–2753.

———, 1964, Trigonometric scales to simplify descriptive geometry in structural geology: Jour. Geological Education, v. 12, p. 126–129.

Travis, R. B., 1964, Geological notes, apparent dip calculator: Am. Assoc. Petroleum Geologists Bull., v. 48, p. 503–504.

Weisbord, N. E., 1935, Graphic method for determination of true dip in pits: Am. Assoc. Petroleum Geologists Bull., v. 19, p. 908–911.

Woolnough, W. G., 1935, Simplification of the John L. Rich dip construction: Am. Assoc. Petroleum Geologists Bull., v. 19, p. 903–908.

CHAPTER 2

Structure Sections

WHILE STUDYING THE STRUCTURES OF AN AREA, and particularly when reporting on them, a geologist must be able to represent their arrangement effectively. Because three-dimensional models are bulky to store and their construction is difficult, slow, and expensive, it is usually more practical to represent structures by means of two-dimensional drawings such as maps and sections. A *geologic map* shows structures by means of formational outcrop patterns. A diagram representing a vertical slice through the upper part of the lithosphere and showing the vertical dimension plus one horizontal dimension is called a *cross section* or *structure section.*

Occasionally one will want to prepare a structure section in some plane other than a vertical one. An example would be a section of the structure as it intersects the footwall of an inclined fault. Such an inclined structure section is beyond the scope of the present chapter, but the procedure should be evident from a study of this chapter combined with the projection techniques described in Chapter 9.

Vertical structure sections can be constructed entirely from maps, from field data alone, or from combinations of these. The preparation of all structure sections begins with the construction of a topographic profile along the line of section. The *line of section* is the line formed by the intersection of the vertical plane of the structure section with the earth's surface. The topographic profile can be drawn from a topographic map, but if none is available, it may be necessary to survey a topographic profile along the line of section and plot the data obtained. Complex equipment also permits the use of aerial photographs to prepare a topographic profile.

A geologic map is often available superimposed on a topographic map base, in which case a profile can be drawn from the combined map. If a separate topographic map with the same scale can be found, it is easier to construct the surface profile from that map and then plot the geologic information from the geologic map. When geologic and topographic maps with the same scale cannot be obtained, one can be changed to coincide with the scale of the other by the methods of enlargement or reduction of scale shown on pages 24–28.

Basic Construction of Structure Sections

The following procedure is suggested for drawing structure sections from geologic and topographic maps with the same scales, if you wish to prepare the structure section with the same horizontal scale as the maps (see Fig. 2–1).

1. Mark a horizontal line on cross-section paper or some other ruled paper to represent the base line of the structure section, and label it according to the altitude it represents. Draw a line extending upward from and perpendicular to one end of the base line. The desired vertical scale should be marked on this vertical line, which is one end of the structure section. If paper with printed ruling is not used, draw lines through the calibrated points on the vertical scale and parallel to the base line to serve as lines of equal altitude to aid the plotting of altitudes on the topographic profile. The horizontal scale can be taken directly from the map, if the section is to be the same scale as the map. (The horizontal scale can be enlarged or reduced also, using methods described later in this chapter.)

2. Place the base line of the structure section so that it is parallel to the line of the section on the map, with the projected end of the structure section passing through the corresponding end of the section line on the map.

3. By placing one side of a right triangle on the base line and passing the perpendicular side of the triangle through the desired point on the map (where a contour line crosses the line of section), that map point can be transferred to its proper hori-

FIG. 2–1 Preparation of topographic profile, with no alteration of horizontal scale.

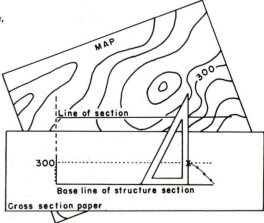

zontal coordinate on the structure section. The vertical coordinate of the map point is located according to the vertical scale at the designated elevation above the base line of the cross section, as at X in Figure 2–1. By repeating this process for each map contour line, a series of points is made on the structure section paper corresponding to altitudes (contour intersections) along the line of section. A line passing through this series of points forms the topographic profile. Plotting is facilitated if a straight edge is placed parallel to the base line and the horizontal edge of the triangle is kept firmly against the straight edge, thus ensuring that the triangle is "squared" at all times.

4. Formation contact lines are marked in their proper positions on the topographic profile by the same method. The dip angle should be plotted to a short distance below the ground surface. The section should be extended to the farthest depth at which the structure can be confidently interpreted, generally not exceeding a mile. Of course, schematic sections may extend to much greater depths if desired. Some areas are so complex that reliable sections can be extended only a few thousand feet beneath the topographic surface. If the dip direction is parallel to the line of section, the dip amount on the cross section is the same as that marked on the map. If the dip direction differs from the direction of the line of section, the apparent dip in the cross section is less than the true dip (see Chapter 1 to calculate the correction). If the true and apparent dip directions do not differ by more than 10°–15°, the dip correction is usually so slight that, for practical purposes, the true dip amount can be used in the structure section. Faults and axes of folds should be plotted on the cross section. After the surface positions of the beds are plotted, then their subsurface positions should be sketched from the available data, keeping formation contacts parallel unless definite depositional or tectonic lensing of the beds is known (see Fig. 2–2).

The method of drawing a structural interpretation at depth depends on whether the folds are concentric folds of competent beds, flow folding of incompetent strata, shear folding, or folding related to thinning of the strata by compaction. In a region with simple structure the assumption of concentric folding usually is valid, and the folds are drawn as concentric arcs, keeping the thickness of individual beds constant. Precise methods for drawing concentric folds are presented by Busk (1929) and Badgley (1959, pp. 32–36). In extreme cases of flow folding, tectonic thinning and swelling of bed thickness can be portrayed best by a combination of free-hand drawing and use of the concept of thinned strata on the limbs of folds, with thickening in the axes. For sections involving incompetent and competent beds that have undergone partial compaction rather than plastic flow, the "boundary ray" method yields consistent results (Coates, 1945; Gill, 1953; and Badgley, 1959, pp. 37–50). Severely deformed sections with alternating competent and incompetent strata are best drawn with concentric folds in the competent beds and flow folds or similar folds in the incompetent layers.

5. The structure section should include at least the following information, which should be centered preferably above the profile: a title consisting of location of selection,

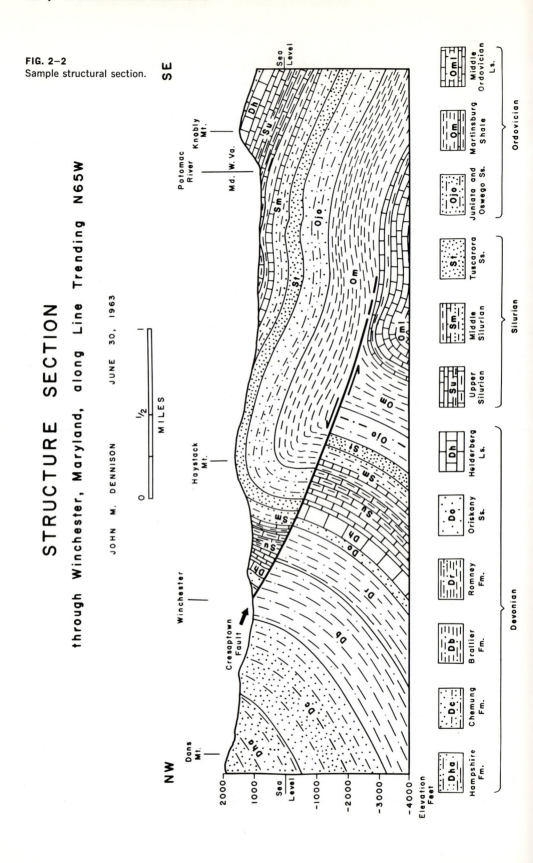

FIG. 2–2
Sample structural section.

date prepared, and name of geologist preparing section. Horizontal and vertical scales should be shown using bar scales. The section itself should show identification of important topographic landmarks, labels for all major faults and folds, designation of relative movements along faults, and identification of formations and lithologies. Commonly below the base line there is shown a legend of beds in order of correct age relationships. Their lithologies are indicated by conventional patterns in the rectangular forms. Careful inking and coloring of the formations and their equivalent legend boxes usually completes the section (Fig. 2–2).

Vertical Exaggeration of Scale

It is sometimes useful to have a larger vertical than horizontal scale in a structure section. This might be desirable in the following situations: if the relief along the line of section is very small, if it is wished to accentuate the relation between topographic irregularities and structure, or if the dips of the beds are very gentle.

In such cases it is essential to know the amount of vertical exaggeration, and this value should be labeled on the structure section. The amount of vertical exaggeration can be calculated from the formula

$$\text{Vertical exaggeration} = \frac{\text{Representative fraction of vertical scale}}{\text{Representative fraction of horizontal scale}}. \tag{4}$$

For example, if the horizontal scale is 1:62,500 and the vertical scale is to be one inch equals 1,000 feet, then

$$\text{Vertical exaggeration} = \frac{\dfrac{1}{12,000}}{\dfrac{1}{62,500}} = \frac{62,500}{12,000} = 5.21 .$$

If the vertical scale of a profile is exaggerated, the ground slopes appear steeper than they actually are on the earth's surface (see Fig. 2–3). The dip angles of beds, faults, and other lines in the structure section must be similarly exaggerated.

The necessary amount of dip exaggeration can be obtained by two methods:

A. Using a graphic method as in Figure 2–3, draw a straight line ABC and divide it so that:

$$\frac{AC}{AB} = \frac{\text{Representative fraction of vertical scale}}{\text{Representative fraction of horizontal scale}} = \text{Vertical exaggeration.} \tag{5}$$

At B set off an angle ABX that is equal to the complement of the unexaggerated dip angle, and let this be met by a line passing through A and perpendicular to line ABC. Draw a line XC passing from C to the intersection of AX and BX. The angle AXC is the exaggerated dip angle to use in the structure section. Note also that angle AXB is the unexaggerated dip angle. In a hypothetical case in which the vertical exaggeration is 5.21 and the unexaggerated dip angle is 30°, then AC equals 5.21, AB equals 1.00, angle ABX equals 60°, and angle AXC equals approximately 71°.

B. From Figure 2–3 the following equation can be derived:

Tangent of exaggerated dip = (Vertical exaggeration)(Tangent of unexaggerated dip). (6)

The derivation is simple.

AC = (Vertical exaggeration)(AB).

If AX has an arbitrary length of one, then AC has a length equal to the tangent of the exaggerated dip, and the length of AB equals the tangent of the unexaggerated dip angle.

Therefore,

Tangent of exaggerated dip = (Vertical exaggeration)(Tangent of unexaggerated dip).

The following solution is obtained for the hypothetical example used in method A.

Tangent of exaggerated dip = (5.21)(0.577) = 3.006.

Therefore the exaggerated dip is approximately 71°.

To produce the reference graph of Figure 2–4, Formula 6 has been solved for a variety of vertical exaggeration factors and unexaggerated dip angles, yielding exaggerated dip angles whose arctangents were determined from trigonometric tables. The family of curved lines in Figure 2–4 represents the unexaggerated dip, and the rectangular coordinates are as labeled. To illustrate use of the graph, a 5.21 vertical exaggeration of a 30° dip is about 71°. In this example the graph must be interpolated between plotted values of vertical exaggeration. The graph can be read accurately enough for most structural section construction, although for very accurate plotting one may wish to make precise trigonometric calculations.

Figure 2–4 can also be used to determine the true topographic slope angle from a topographic profile with an exaggerated vertical scale.

It should always be remembered that vertically exaggerated structure sections do not furnish a picture in proper proportions of actual conditions in the earth. From vertically exaggerated sections precise measurements cannot be made for tunneling, mining, drilling, or excavation. Structure sections with the same horizontal and vertical scales are preferred for most careful geologic work. When 1:125,000 or 1:62,500 scale maps were used for most geologic reports (prior to the 1950's), it was customary to use a topographic exaggeration in structure sections tens of miles long. For modern work with a 1:24,000 topographic base, it is best to prepare sections with no vertical exaggeration.

Exercises / The purpose of the following exercises is to give the student practice in the general procedure of constructing structure sections and particularly to give experience in preparing a structure section with a vertical exaggeration. Almost any set of geologic and topographic maps of an area can be used for such training. The exercises presented are considered typical. The maps used are from the Hollidaysburg-Huntingdon, Pennsylvania, folio of the Geologic Atlas of the United States, folio No. 227 (Butts, 1945).

The structure section should be compiled in pencil in the conventional manner, including indication of formation lithologies with standard rock patterns. The persistent

FIG. 2–3 Calculation of exaggerated dip.

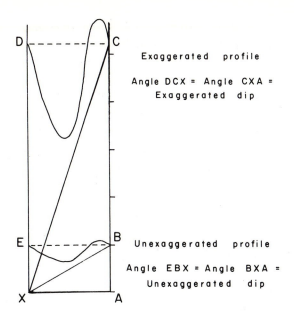

Exaggerated profile

Angle DCX = Angle CXA =
Exaggerated dip

Unexaggerated profile

Angle EBX = Angle BXA =
Unexaggerated dip

FIG. 2–4 Graph for determining exaggerated dips in structure sections. Curved family of lines shows unexaggerated dip angles.

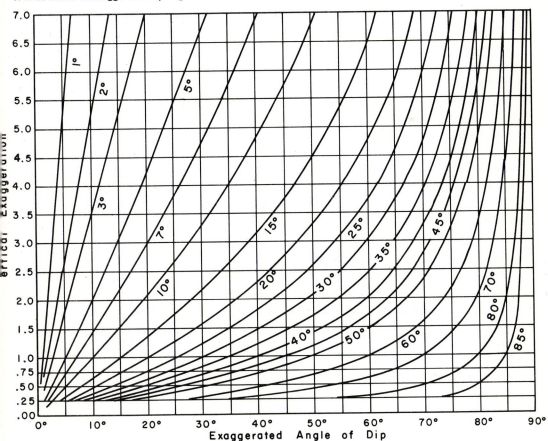

lines of these patterns should be considered analogous to bedding planes and should be drawn parallel to the bedding in anticlines and synclines as shown in Figure 2–2. Plate III of *The Preparation of Illustrations for Reports of the United States Geological Survey* (Ridgway, 1920) shows many of the standard lithologic patterns used. The vertical scale should be 1:12,000 (one inch equals 1,000 feet), and the horizontal scale 1:62,500. Let the base line of the section be 500 feet below sea level. Construct the section from the viewpoint of an observer who is looking northeast.

38. Using the areal geologic map of the Huntingdon, Pennsylvania, quadrangle (Butts, 1945), draw a structure section extending from the northwest corner of the map to the center of the "o" in "Grafton." Be alert for stratigraphic changes within the Clinton Formation.

39. Using the areal geologic map of the Hollidaysburg, Pennsylvania, quadrangle (Butts, 1945), draw a structure section that extends 9.00 miles northwest from latitude 40°15′, longitude 78°20′ and parallel to line of section F-F′. Be alert for overturned beds not marked as such.

Alteration of Horizontal Scale

A geologist is often required to enlarge or reduce a map scale or the horizontal scale of a structure section. The most common situation arises when one wishes to enlarge the scale of a geologic map in order to prepare a section showing structural details. On the other hand, it is sometimes necessary to reduce a map scale when the diagram is to be included in a report or publication in which the size of the map or section is restricted to limits set by the publisher or employer.

The following are among the most common methods of enlarging or reducing horizontal scales.

Engineer's Scale / When other instruments are not available, it is possible to enlarge or reduce a distance on a map by multiplying the map distance measured in inches on an engineer's scale by the enlargement or reduction factor. This factor is the new scale shown as a representative fraction, divided by the representative fraction of the original scale. Note that in the case of enlargement, this factor is greater than one, but it is smaller than one in the case of reduction. If the distance between two points on a map is measured as 1.60 inches, in order to produce a distance enlarged 5.2 times, it would be necessary to plot that distance as 5.2 times 1.60 inches, or 8.32 inches.

The graph in Figure 2–5 facilitates these determinations. It shows the distance to plot on a structure section if the map distance is to be altered by a specified enlargement factor.

Ordinary Drafting Dividers / When a pair of ordinary dividers is available, these may be used to enlarge a scale, although reduction by this method is not practical. Enlargement is accomplished by placing the dividers (with the distance between

FIG. 2–5 Conversion of map distances to structure section distances, for different enlargement factors. Map distance times enlargement factor (diagonal lines) equals structure section distance.

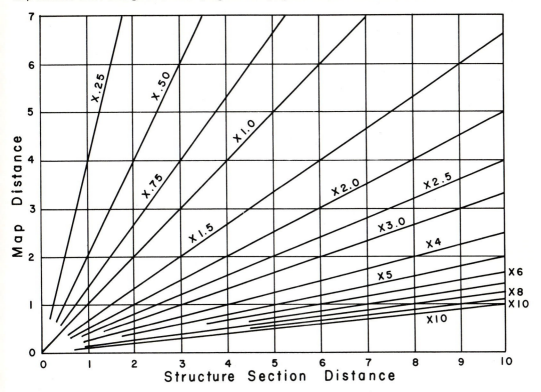

FIG. 2–6 Use of dividers for scale enlargement.

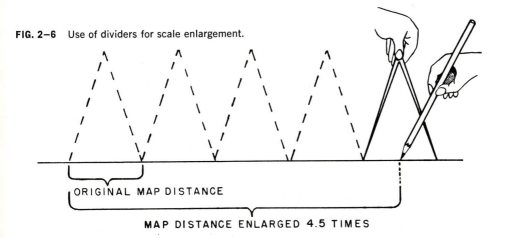

ORIGINAL MAP DISTANCE

MAP DISTANCE ENLARGED 4.5 TIMES

points representing a distance on the original map) along the cross section a successive number of times equal to the enlargement factor. Fractions are interpolated between the last two points as shown in Figure 2–6.

Proportional Dividers / Proportional dividers (Figure 2–7) are manufactured expressly for the enlargement and reduction of scales. They consist of two crossed arms with sharp points at both ends of each arm and with a pivot adjustable lengthwise at the intersection of the two arms. The dividers are designed so that the position of the pivot governs the ratio of the distances between the two pairs of adjacent points on opposite ends of the instrument. This ratio (enlargement or reduction factor) is indicated on the instrument. When the pivot is set at the desired ratio, a map distance laid off at one end of the proportional dividers is automatically enlarged or reduced at the other end of the dividers.

FIG. 2–7 Proportional dividers.

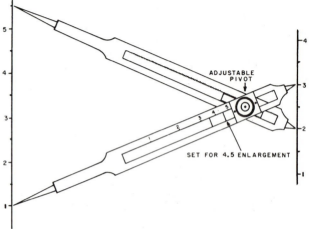

Photographic Methods / Maps or structure section scales can be very effectively reduced photographically. This method has the advantage that imperfections such as wavy lines are reduced so that they are not so apparent as in the original. The end result is a neater illustration. Of course, in any structure section reduced by this manner, the vertical and horizontal scales are reduced by the same factor. It is undesirable to enlarge structure sections photographically, because imperfections become more conspicuous. It is possible to enlarge the original map to the desired scale photographically, and then construct the structure section directly from this enlarged map.

If a drawing is to be reduced in scale, the original drawing should be prepared 1.5 to 4 times the intended final scale. Irregularities of inked line width or slightly irregular lines resulting from an unsteady draftsman become less obvious after reduction. Certain types of imperfections become more obvious after reduction. These include lack of parallelism of lines which should be exactly parallel and a lack of symmetry in placing the component parts of the total illustration. The eye can scan the total illustration at one glance in reduced form, rather than have to examine different parts of the whole illustration in the original larger form.

Copy that is to be enlarged to final form enlarges any errors in the original geologic material, and wiggly lines drawn by an unsteady hand become very obvious. Great care must be taken in drafting illustrations that must be enlarged for final presentation. The most common situation is in preparing copy for slides to be used with a talk.

Geometric Methods / When special instruments are not available, it is feasible to enlarge or reduce map distances by mechanical methods. The procedure is a variation of that used by the cartographer to divide a line into a number of equal parts. It is based on the principle that known distances on one side of a right triangle are enlarged when they are projected perpendicular to that side onto the hypotenuse. The ratio of the lengths of the hypotenuse to the side determines the enlargement factor.

Fold one end of the structure section paper at the desired angle with the base line (Fig. 2–8), and place this folded edge of the section paper coincident with the line of the section on the map (FC). The fold should be made far enough from the end line of the section position so that the folded portion can be cut off the end of the section and not disfigure the completed drawing. The angle FAB determines the enlargement factor of map distances (FC) projected perpendicular to the line of section onto the base line (EG). Map points projected to the base line are placed at the desired elevation along a line perpendicular to the base line by means of a vertical scale. Point H on the structure section represents the proper elevation plot for point F on the line of section, and K represents the plot for C. This procedure is repeated until all contours are represented by points. Connection of these points completes the profile. Formation contacts and faults are plotted in the same manner.

The magnitude of angle FAB of Figure 2–8 is determined by the relationship:

$$\text{Cos FAB} = \frac{AF}{AG} = \frac{\dfrac{\text{Map scale}}{\text{representative fraction}}}{\dfrac{\text{Cross-section horizontal}}{\text{scale representative fraction}}} = \frac{1}{\text{Enlargement factor}}$$

or,

Sec FAB = Enlargement factor.

If the desired cross-section scale is 1:6,000 and the map scale is 1:12,000, then

$$\text{Cos FAB} = \frac{1/12,000}{1/6,000} = .500 \ .$$

Angle FAB = 60°.

The scale of a map can be reduced in drawing a structure section if the relative positions of the map and cross section in Figure 2–8 are reversed and if points are projected from the line of section on the map perpendicular to the base line of the cross section.

FIG. 2–8 Horizontal scale enlargement by geometric method.

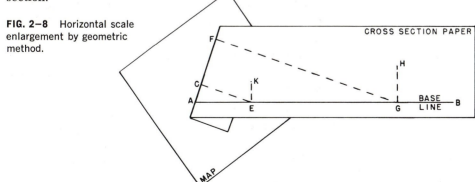

Mechanical alteration of scales can be simplified if a 4-foot or longer T-square and a table with a straight edge are available. Place the line of section (on the map) parallel to the table edge. A T-square with the head "squared" along the table edge automatically projects lines perpendicular to the line of section onto the base line. The length of the section (EG in Fig. 2–8) is calculated by multiplying the enlargement factor by the length of the line of section on the map (FC). The extremities of the section should be marked on both the map and the section paper. The angle between the base line and the line of section is then fixed by adjusting the positions of the lines until the ends of the structure section intersect with the corresponding ends of the line of section on the map as projected by the T-square. This obviates the calculation of angle FAB and eliminates the necessity of folding the structure section paper. The map and cross-section paper are fastened in place (with *drafting tape*—not transparent tape, which ruins the maps), the map points are mechanically transferred to their correct positions on the section by means of the T-square, and the topographic profile is constructed from these points. Formation contacts and faults are plotted in the same manner.

Exercises / The following are suggested as practical exercises in constructing structure sections with larger scales than the maps from which they are drawn. Unless otherwise specified, place the base of the section at 1,000 feet below sea level. Indicate formation lithologies and other features as recommended on pages 19–21. Ink and color the section. Neatness is imperative. Check for content with Figure 2–2.

The exercises are listed by categories of structural situations. In general the easiest structural sections are listed first.

Most of the exercises are based on U.S. Geological Survey maps of the Geologic Quadrangle Map series (GQ series). A few exercises are based on maps published by state geologic surveys.

[Nearly Flat Strata to Gentle Folds]

40. Amity Quadrangle, Pennsylvania, by H. L. Berryhill, Jr. (1964). U.S. Geol. Sur. GQ-296; scale 1:24,000. Draw a structure section as viewed facing north and extending from the northwest to the southeast corner of the quadrangle. Use a vertical scale of 1:8,000 and a horizontal scale of 1:24,000.

41. Epes Quadrangle, Alabama, by W. H. Monroe and J. L. Hunt (1958). U.S. Geol. Sur. GQ-113: scale 1:62,500. Draw a structure section extending along the 88°10′ meridian, viewed facing to the west. Use a horizontal scale of 1:31,250 and exaggerate the vertical scale 6 times over the horizontal scale.

42. Mammoth Cave Quadrangle, Kentucky, by D. D. Haynes (1964). U.S. Geol. Sur. GQ-351; scale 1:24,000. Draw a structure section as viewed facing north and extending from the northwest to the southeast corner of the quadrangle. Use a horizontal scale of 1:24,000 and a vertical scale of 1:8,000.

43. Trenton East Quadrangle, New Jersey-Pennsylvania (Pre-Quaternary Geology), by J. P Owens and J. P. Minard (1964). U.S. Geol. Sur. GQ-341; scale 1:24,000. Draw a structure section as viewed facing north and extending from the northwest

to the southeast corner of the quadrangle. Use a horizontal scale of 1:24,000 and a vertical exaggeration of 10 times.

[Nearly Flat Strata with Simple Faults]

44. Pleasant Green Hill Quadrangle, Kentucky, by W. H. Nelson (1964). U.S. Geol. Sur. GQ-321; scale 1:24,000. Draw a structure section as viewed facing northwest and extending from the southwest to the northeast corner of the quadrangle. Use a 1:24,000 horizontal scale and a vertical scale of 1:8,000.

45. Salem Quadrangle, Kentucky, by R. D. Trace (1962). U.S. Geol. Sur. GQ-206; scale 1:24,000. Draw a structure section viewed facing north and extending east-west along latitude 37°18′N. Use a scale of 1:24,000.

46. Adam Weiss Peak Quadrangle, Wyoming, by W. L. Rohrer (1965). U.S. Geol. Sur. GQ-382; scale 1:24,000. Draw a structure section along the west border of the quadrangle with no vertical exaggeration, using a scale of 1:24,000. Draw the section with the north end on the right.

[Folded Strata]

47. Mifflintown Quadrangle, Pennsylvania, by R. R. Conlin and D. M. Hoskins (1962). Geologic Atlas of Pennsylvania, Atlas 126; Pennsylvania Topographic and Geologic Survey; scale 1:24,000. Draw a structure section extending from 40°42′30″N, 77°30′W to 40°30′N, 77°22′30″W with no vertical exaggeration. Draw the section as viewed facing northeast, using a scale of 1:24,000.

48. Pysht Quadrangle, Washington, by H. D. Gower (1960). U.S. Geol. Sur. GQ-129; scale 1:62,500. Draw a structure section along meridian 124°10′W, extending from the Soleduck River to 48°15′N latitude. Use a vertical scale of 1:31,250 with no vertical exaggeration. Place the north end of the section on the right, and draw the section to a depth of 2,000 feet below sea level.

49. Bedford Quadrangle, Wyoming, by W. W. Rubey (1958). U.S. Geol. Sur. GQ-109; scale 1:62,500. Draw a section with a scale of 1:20,000 extending from the section corner within the town of Grover due east to the eastern margin of the quadrangle. Draw the section as viewed facing north.

[Folded and Faulted Strata]

50. Martinsburg Quadrangle, West Virginia, by R. C. Page and others (1964). West Virginia Geological Survey, Geologic Map GM-2; scale 1:24,000. Draw a scale 1:24,000 section extending from 39°30′N, 78°0′W to 39°27′30″N, 77°52′30″W. Draw the section as viewed facing north.

51. Sugar House Quadrangle, Utah, by M. D. Crittenden, Jr. (1965). U.S. Geol. Sur. GQ-380; scale 1:24,000. Draw a structure section with a scale 1:24,000 extending along meridian 111°46′0″W. Place the north end of the section on the left.

52. Boulter Peak Quadrangle, Utah, by A. E. Drisbow (1961). U.S. Geol. Sur. GQ-141; scale 1:24,000. Draw an east-west structure section at 40°1′30″N latitude, extending completely across the quadrangle. Use a scale of 1:24,000 and extend the section as deeply into the ground as you feel confident. Draw the section as viewed facing north.

53. Brisbin Quadrangle, Montana, by A. E. Roberts (1964). U.S. Geol. Sur. GQ-256;

scale 1:24,000. Draw a 1:15,000 scale structure section along the east margin of the quadrangle. Place the north end of the section on the right, and extend the section to a depth of 1,000 feet above sea level.

[Gently Deformed Strata with Unconformity]

54. Calamity Mesa Quadrangle, Colorado, by F. W. Carter, Jr. (1955). U.S. Geol. Sur. GQ-61; scale 1:24,000. Draw a structure section from the southwest to the northeast corner of the quadrangle, using a vertical and horizontal scale of 1:24,000. Place the northeast end of the section on the right, and the base of the section at 5,000 feet above sea level.

55. Miners Delight Quadrangle, Wyoming, by R. W. Bayley (1965). U.S. Geol. Sur. GQ-460; scale 1:24,000. Draw a structure section along meridian 108°41'15"W, using a scale of 1:12,000. Place the north end of the section on the right, and extend the section downward to an elevation of 7,000 feet above sea level.

[Folds, Faults, and Unconformity]

56. Anderson Mesa Quadrangle, Colorado, by F. W. Carter, Jr. (1955). U.S. Geol. Sur. GQ-77; scale 1:24,000. Draw a 1:12,000 scale structural section along latitude 38°11'0"N. Extend the section to a depth of 5,000 feet above sea level. Place the east end of the section on the right.

[Thrust Faults]

57. Athens Quadrangle, Tennessee, by J. Rodgers (1952). U.S. Geol. Sur. GQ-19; scale 1:24,000. Draw a structure section with a scale of 1:24,000 extending from 35°28'N, 84°37'30"W to 35°24'N, 84°30'W. Draw the section as viewed facing north, and place the base of the section at 3,000 feet below sea level.

58. Rockwood Quadrangle, Tennessee, by G. D. Swingle (1960). Tennessee Division of Geology GM-123-SW; scale 1:31,680. Draw a structural section along the east margin of the quadrangle, using a scale of 1:24,000. Place the north end of the section on the right, and extend the section to a depth of 2,000 feet below sea level.

[Igneous Intrusion with Simply Deformed Strata and Unconformities]

59. Timpanogos Cave Quadrangle, Utah, by A. A. Baker and M. D. Crittenden, Jr. (1961). U.S. Geol. Sur. GQ-132; scale 1:24,000. Draw a structure section with a scale of 1:24,000 along meridian 111°38'30"W. Extend the section down to an elevation of 5,000 feet above sea level. Place the north end of the section on the left.

[Folded and Faulted Strata with Igneous Intrusions]

60. Packsaddle Mountain Quadrangle, Idaho, by J. E. Harrison and D. A. Jobin (1965). U.S. Geol. Sur. GQ-375; scale 1:62,500. Draw a 1:20,000 scale structural section along latitude 48°4'30"N, extending from the east shore of Pend Oreille Lake to the east margin of the quadrangle. Place the east end of the section on the right.

61. Chimney Rock Quadrangle, Montana, by A. E. Roberts (1964). U.S. Geol. Sur. GQ-257; scale 1:24,000. Draw a 1:15,000 scale structure section extending from 45°31'30"N, 110°45'W to 45°37'30"N, 110°41'W. Extend the section to a depth of 4,000 feet above sea level, and place the north end of the section on the right.

[Volcanics]

62. Harvey Mountain Quadrangle, California, by G. A. Macdonald (1965). U.S. Geol.

Sur. GQ-443; scale 1:62,500. Draw a 1:31,250 scale structure section from 40°30′N, 121°2′W to 40°45′N, 121°8′W. Draw the base of the section at 5,000 feet above sea level, and place the north end of the section on the left.

63. Prospect Peak Quadrangle, California, by G. A. Macdonald (1964). U.S. Geol. Sur. GQ-345; scale 1:62,500. Draw a structure section with a scale of 1:31,250 along meridian 121°21′W. Place the north end of the section on the left, and make the base of the section 3,000 feet above sea level.

64. Cameron Quadrangle, Arizona, by J. P. Akers and others (1962). U.S. Geol. Sur. GQ-162; scale 1:62,500. Draw a structure section extending from 36°0′N, 111°30′W to 35°57′N, 111°15′W. Use a horizontal scale of 1:62,500 and a vertical scale of 1:31,250. Extend the section down to sea level. Draw the section as viewed facing north.

[Volcanics and Sedimentary Rocks]

65. Avon Quadrangle, Connecticut (Bedrock Geology), by R. W. Schnabel (1960). U.S. Geol. Sur. GQ-134; scale 1:24,000. Draw a 1:24,000 scale section along latitude 41°47′N. Place the east end of the section on the right, and extend the section to a depth of 2,000 feet below sea level, where sufficient information is available.

66. Ouray Quadrangle, Colorado, by R. G. Luedke and W. S. Burbank (1962). U.S. Geol. Sur. GQ-152; scale 1:24,000. Draw a 1:24,000 scale section from 38°1′0″N, 107°45′0″W to 38°7′30″N, 107°37′30″W. Draw the section as viewed facing north, and place the base of the section at elevation 7,000 feet above sea level.

[Folded and Faulted Sediments, Metamorphic and Igneous Rocks]

67. Hayward Quadrangle, California, by G. D. Robinson (1956). U.S. Geol. Sur. GQ-88; scale 1:24,000. Draw a 1:12,000 scale structure section from 37°41′0″N, 122°5′15″W to 37°43′0″N, 122°0′0″W. Draw the section as viewed facing north.

68. Paradise Peak Quadrangle, Nevada, by C. J. Vitaliano and E. Callaghan (1963). U.S. Geol. Sur. GQ-250; scale 1:62,500. Draw a 1:20,000 scale structure section from 38°49′0″N, 117°56′0″W to 38°55′30″N, 117°45′0″W. Extend section downward to elevation 4,000 feet. Draw the section as viewed facing north.

[Metamorphic and Igneous Terrain]

69. Danforth Quadrangle, Maine, by D. M. Larrabee and C. W. Spencer (1963). U.S. Geol. Sur. GQ-221; scale 1:62,500. Draw a 1:31,250 scale structure section extending from 45°30′N, 68°0′W to 45°39′N, 67°45′W. Place the east end of the section on the right. Extend the section to a depth of a mile below sea level.

70. Greenville Quadrangle, Maine, by G. H. Espenshade and E. L. Boudette (1964). U.S. Geol. Sur. GQ-330; scale 1:62,500. Draw a structure section extending from 45°15′N, 69°30′W to 45°25′N, 69°45′W, using a scale of 1:30,000 for both the vertical and horizontal dimension. Draw the section as viewed facing north, and place the base of the section at 2,000 feet below sea level.

71. Willimantic Quadrangle, Connecticut, by G. L. Snyder. U.S. Geol. Sur. GQ-335; scale 1:24,000. Draw a 1:24,000 scale structure section from the northwest to the southeast corner of the quadrangle. Place the northwest end of the section on the left.

72. Lincoln Mountain Quadrangle, Vermont, by W. M. Cady and others (1962). U.S. Geol. Sur. GQ-164; scale 1:62,500. Draw a 1:30,000 scale cross section along latitude 44°3'N, viewed facing northward.

73. North Adams Quadrangle, Massachusetts-Vermont, by N. Herz (1961). U.S. Geol. Sur. GQ-139; scale 1:24,000. Draw an east-west structural section completely across the quadrangle extending along the Hoosac Tunnel. Extend the section to 1,500 feet below sea level depth. Show the level of the approximately horizontal tunnel in the cross section. Place the west edge of the structure section on the left.

References

Adams, G. F., 1957, Block diagrams from perspective grids: Jour. Geological Education, v. 5, no. 2, p. 10–19.

Akers, J. P., and others, 1962, Cameron Quadrangle, Arizona: U.S. Geol. Sur., Map GQ-162.

Badgley, P. C., 1959, Structural methods for the exploration geologist: New York, Harper and Bros., 280 p.

Baker, A. A., and Crittenden, M. D., Jr., 1961, Timpanogos Cave Quadrangle, Utah: U.S. Geol. Sur., Map GQ-132.

Bayley, R. W., 1965, Miners Delight Quadrangle, Wyoming: U.S. Geol. Sur., GQ-460.

Berryhill, H. L., Jr., 1964, Amity Quadrangle, Pennsylvania: U.S. Geol. Sur. GQ-296.

Busk, H. G., 1929, Earth flexures, their geometry and their representation and analysis in geological section with special reference to the problem of oil finding: Cambridge University Press, 106 p. (Reprinted New York, William Trussell, 1957.)

Butts, C., 1945, Hollidaysburg-Huntingdon, Pennsylvania, Folio: U.S. Geol. Sur., Geol. Atlas of the U.S., folio 227, 20 p.

Cady, W. M., and others, 1962, Lincoln Mountain Quadrangle, Vermont: U.S. Geol. Sur., GQ-164.

Carter, F. W., Jr., 1955, Calamity Mesa Quadrangle, Colorado: U.S. Geol. Sur., Map GQ-61.

————, 1955, Anderson Mesa Quadrangle, Colorado: U.S. Geol. Sur., GQ-77.

Chalmers, R. M., 1926, Geological maps, the determination of structural detail: London, Oxford University Press, 175 p.

Coates, J., 1945, The construction of geological sections: Quart. Jour. Geol. Min. Met. Soc. India, v. 17, no. 1.

Conlin, R. R., and Hoskins, D. M., 1962, Mifflintown Quadrangle, Pennsylvania: Geologic Atlas of Pennsylvania, Atlas 126.

Crittenden, M. D., Jr., 1965, Geology of the Dromedary Peak Quadrangle, Utah: U.S. Geol. Sur., Map GQ-378.

————, 1965, Sugar House Quadrangle, Utah: U.S. Geol. Sur., Map GQ-380.

Drisbow, A. E., 1961, Boulter Peak Quadrangle, Utah: U.S. Geol. Sur., Map GQ-141.

Eardley, A. J., 1938, Graphic treatment of folds in three dimensions: Am. Assoc. Petroleum Geologists Bull., v. 22, p. 483–489.

Espenshade, G. H., and Boudette, E. L., 1964, Greenville Quadrangle, Maine: U.S. Geol. Sur., Map GQ-330.

Gill, W. D., 1953, Construction of geological sections of folds with steep limb attenuation: Am. Assoc. Petroleum Geologists Bull., v. 37, p. 2389–2406.

Gower, H. D., 1960, Pysht Quadrangle, Washington: U.S. Geol. Sur., GQ-129.

Harrison, J. E., and Jobin, D. A., 1965, Packsaddle Mountain Quadrangle, Idaho: U.S. Geol. Sur., Map GQ-375.

Haynes, D. D., 1964, Mammoth Cave Quadrangle, Kentucky: U.S. Geol. Sur., Map GQ-351.

Herold, S. C., 1933, Projection of dip angle on profile section: Am. Assoc. Petroleum Geologists Bull., v. 17, p. 740–742.

Herz, N., 1961, North Adams Quadrangle, Massachusetts: U.S. Geol. Sur., Map GQ-139.

Johnston, W. D., and Nolan, T. B., 1937, Isometric block diagrams in mining geology: Econ. Geology, v. 32, p. 550–569.

Knuston, R. M., 1958, Structural sections and the third dimension: Econ. Geology, v. 53, p. 270–286.

Larrabee, D. M., and Spencer, C. W., 1963, Danforth Quadrangle, Maine: U.S. Geol. Sur., Map GQ-221.

LeRoy, L. W., and Low, J. W., 1954, Graphic problems in petroleum geology: New York, Harper and Bros., p. 140–143. (Busk method of drawing folds.)

Lobeck, A. K., 1958, Block diagrams and other graphic methods used in geology and geography: Amherst, Mass., Emerson-Trussell Book Co., 2nd ed., 212 p.

Luedke, R. G., and Burbank, W. S., 1962, Ouray Quadrangle, Colorado: U.S. Geol. Sur., GQ-152.

Macdonald, G. A., 1964, Prospect Peak Quadrangle, California: U.S. Geol. Sur., GQ-345.

———, 1965, Harvey Mountain Quadrangle, California: U.S. Geol. Sur., GQ-443.

Mackin, J. H., 1950, The down-structure method of viewing geologic maps: Jour. Geology, v. 58, p. 55–72.

Malhase, J., 1927, Constructing geologic sections with unequal scales: Am. Assoc. Petroleum Geologists Bull., v. 11, p. 755–757.

Mertie, J. B., 1947, Calculation of thickness in parallel folds: Geol. Soc. America Bull., v. 58, p. 779–802.

———, 1948, Application of Brianchon's Theorem to construction of geologic profiles: Geol. Soc. America Bull., v. 59, p. 767–786.

Monroe, W. H., and Hunt, J. L., 1958, Epes Quadrangle, Alabama: U.S. Geol. Sur., Map GQ-113.

Moran, W. R., editor, 1964, American Association of Petroleum Geologists slide manual: American Association of Petroleum Geologists, Tulsa, Okla., 32 p.

Nelson, W. H., 1964, Pleasant Green Hill Quadrangle, Kentucky: U.S. Geol. Sur., GQ-321.

Owens, J. P., and Minard, J. P., 1964, Trenton East Quadrangle, New Jersey–Pennsylvania Pre-Quaternary Geology): U.S. Geol. Sur., GQ-341.

Page, R. C., and others, 1964, Martinsburg Quadrangle, West Virginia: West Virginia Geol. Sur., Geologic Map GM-2.

Raisz, E., 1962, Principles of cartography: New York, McGraw-Hill Book Co., 315 p.

Ridgway, J. L., 1920, The preparation of illustrations for reports of the United States Geological Survey: U.S. Geol. Sur., 101 p.

Roberts, A. E., 1964, Brisbin Quadrangle, Montana: U.S. Geol. Sur., Map GQ-256.

———, 1964, Chimney Rock Quadrangle, Montana: U.S. Geol. Sur., Map GQ-257.

Robinson, G. D., 1956, Hayward Quadrangle, California: U.S. Geol. Sur., Map GQ-88.

Rodgers, J., 1952, Athens Quadrangle, Tennessee: U.S. Geol. Sur., GQ-19.

Rohrer, W. L., 1965, Adam Weiss Peak Quadrangle, Wyoming: U.S. Geol. Sur., GQ-382.

Rubey, W. W., 1958, Bedford Quadrangle, Wyoming: U.S. Geol. Sur., GQ-109.

Secrist, M. H., 1936, Perspective block diagrams: Econ. Geology, v. 31, p. 867–880.

Schnabel, R. W., 1960, Avon Quadrangle, Connecticut (Bedrock Geology): U.S. Geol. Sur., Map GQ-134.

Snyder, G. L., Willimantic Quadrangle, Connecticut: U.S. Geol. Sur., GQ-335.

Stockwell, C. H., 1950, The use of plunge in the construction of cross-sections of folds: Geol. Assoc. Canada Proc., v. 3, p. 97–121.

Swingle, G. D., 1960, Rockwood Quadrangle, Tennessee: Tennessee Division of Geology, Map GM-123-SW.

Trace, R. D., 1962, Salem Quadrangle, Kentucky: U.S. Geol. Sur., GQ-206.

United States Geological Survey, 1958, Suggestions to authors of the reports of the United States Geological Survey: U.S. Govt. Printing Office, 5th ed., 255 p.

Vitaliano, C. J., and Callaghan, E., 1963, Paradise Peak Quadrangle, Nevada: U.S. Geol. Sur., Map GQ-250.

Wentworth, C. N., 1917, A proposed dip protractor: Jour. Geology, v. 25, p. 489–491.

Woolnough, W. G., and Benson, W. N., 1957, Graphical determination of the dip in deformed and cleaved sedimentary rocks: Jour. Geology, v. 65, p. 428–433.

CHAPTER 3

Thickness Determinations

A GEOLOGIST WORKING OUT STRUCTURES IN THE FIELD must be able to calculate the thickness of formations from data gathered at the outcrop, because he will seldom find all of the formations exposed where their thicknesses can be measured directly by tape.

The *thickness* of a bed or any other tabular form is the distance from the subface (bottom boundary surface) to the superface (top surface) measured along a line normal to these bounding surfaces. The strike, dip, and thickness directions are thus mutually perpendicular lines, producing right-triangle relationships in any outcrop of strata. By relating the outcrop position of the superface and subface to the attitude of these mutually perpendicular lines, the thickness of the bed can be determined either trigonometrically or graphically.

Formation thickness is measured perpendicular to the bedding surface in a vertical cross section parallel to the dip direction and normal to the strike. The outcrop positions of the superface and subface are projected onto this vertical plane normal to the strike, and the true thickness is determined in this plane. In graphic solutions the outcrop relationships are plotted and the thickness is measured in scale drawings. Thickness can also be calculated by trigonometry.

If the ground slope does not change appreciably in magnitude or direction, it may be considered as a plane formed by the line of slope along a traverse and a level line on the ground surface. The thickness of the bed can be calculated from the dip of the bed and the traverse direction along with one of the following groups of information:

1. the traverse distance and the angle of ground slope along the traverse,

2. the traverse distance and change of altitude between the outcrops of superface and subface, or
3. the horizontal (map) distance and change in altitude between the outcrops of the superface and subface.

If the slope changes significantly in direction or magnitude, the horizontal (map) distance and the difference in altitude between the outcrops of the superface and subface must be used to calculate the bed thickness.

The following outline presents different types of solutions for calculating bed thickness. Illustrated solutions are given, along with derivations of trigonometric formulae. For graphical determinations the relationships are drawn to scale and measured in the same sequence as the steps of the trigonometric calculations.

Simple Cases

Horizontal Beds / The thickness of a horizontal bed is the difference in altitude between the superface and subface as determined by use of a hand level, aneroid barometer, topographic map, or surveying methods.

Vertical Beds / The thickness of a vertical bed is the horizontal distance between the superface and subface measured in a direction perpendicular to the strike. On level ground this can usually be measured directly. The following formulae can be used if the ground surface slopes and the locations of a point on the superface and on the subface are known (see Fig. 3–1):

If the traverse is not perpendicular to the strike,

$$\text{Thickness} = \text{(Traverse distance between superface and subface)(Cos traverse slope angle)(Sin angle between traverse and strike directions).} \tag{7}$$

$$\text{Thickness} = \text{(Traverse distance)} \left[\text{Cos arc sin} \left(\frac{\text{Difference in altitude along traverse}}{\text{Traverse distance}} \right) \right] \text{(Sin angle between strike and traverse directions).} \tag{8}$$

$$\text{Thickness} = \text{(Horizontal distance between superface and subface)(Sin angle between traverse and strike directions).} \tag{9}$$

$$\text{Thickness} = \text{(Sin angle between strike and traverse directions)} \left[\sqrt{\left(\frac{\text{Traverse}}{\text{distance}} \right)^2 - \left(\frac{\text{Difference in altitude between}}{\text{superface and subface outcrops}} \right)^2} \right]. \tag{10}$$

If the traverse is perpendicular to the strike, then these formulae simplify to:

$$\text{Thickness} = \sqrt{\left(\frac{\text{Traverse}}{\text{distance}} \right)^2 - \left(\frac{\text{Difference in altitude between}}{\text{superface and subface outcrops}} \right)^2}. \tag{11}$$

$$\text{Thickness} = \text{(Traverse distance)(Cos traverse slope angle).} \tag{12}$$

A graphical solution to the problem diagrammed in Figure 3–1 is presented in the scale drawing of Figure 3–2.

Traverse Perpendicular to Strike of Dipping Beds

A. Level ground surface.

    Thickness = (Traverse distance) (Sin dip angle).   **(13)**

B. Sloping ground surface.

1. Ground surface slopes in dip direction.

    The following formula can be derived from Figure 3–3.

    Thickness = AB sin ABF.

    Thickness = AB sin (JBF − JBA).

    Thickness = (Traverse distance) [Sin (dip angle − slope angle)].   **(14)**

Frequently the traverse distance and the difference in altitude along the slope are determined, in which case (Figure 3–3):

    Thickness = AG − FG, = AG − HC.

    Thickness = AC(Sin ACG) − BC(Cos BCH).

    Thickness = $\left(\text{AB Cos arc sin }\dfrac{\text{BC}}{\text{AB}}\right)$ (Sin ACG) − BC(Cos ACG).

    Thickness = (Traverse distance) $\left(\text{Cos arc sin }\dfrac{\textbf{Difference in altitude}}{\textbf{Traverse distance}}\right)$ (Sin

    dip angle) − (Difference in altitude)(Cos dip angle).   **(15)**

The data used may be horizontal distance from superface to subface outcrops measured perpendicular to strike (for example, taken from a map) along with the difference in elevation between these two points. In that case (Fig. 3–3):

    Thickness = AC(Sin ACG) − BC(Cos HCB).

    Thickness = AC(Sin ACG) − BC(Cos ACG).

    Thickness = (Horizontal distance)(Sin dip angle) − (Difference in altitude)(Cos dip angle).   **(16)**

2. Ground surface slopes opposite to dip direction.

    See Figure 3–4 for illustration of this situation. Formulae can be derived for solving this type of problem in a manner similar to those developed from Figure 3–3.

    Thickness = (Traverse distance) [Sin (dip angle + slope angle)].   **(17)**

    Thickness = (Traverse distance $\left(\text{Cos arc sin }\dfrac{\textbf{Difference in altitude}}{\textbf{Traverse distance}}\right)$ (Sin

    dip angle) + (Difference in altitude)(Cos dip angle).   **(18)**

    Thickness = (Horizontal distance)(Sin dip angle) + (Difference in altitude)(Cos dip angle).   **(19)**

Exercises / Exercises 74–93 concern the preceding types of formation thickness problems. Work out both graphic and trigonometric solutions. Answers expressed in three significant figures are sufficient considering the probable accuracy of these types of data. It is very difficult to determine strike and dip angles in the field with an accuracy better than 1°.

    In actual practice in the field, it is very convenient to record the necessary data,

FIG. 3-1 Calculation of thickness on sloping terrain with vertical bedding.

FIG. 3-2 Graphical solution to stratigraphic thickness problem diagrammed in Figure 3-1.

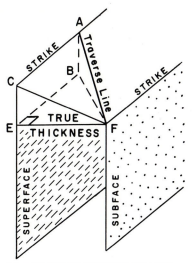

AF — TRAVERSE DISTANCE OBLIQUE TO STRIKE.
CF — SLOPE DISTANCE PERPENDICULAR TO STRIKE
BF — HORIZONTAL DISTANCE OBLIQUE TO STRIKE
EF — TRUE THICKNESS

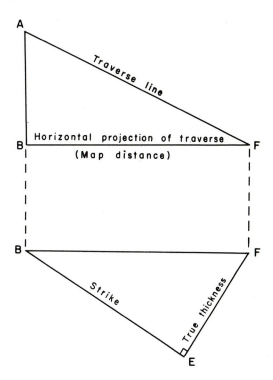

FIG. 3-3 Thickness calculation where ground slopes in dip direction, with traverse perpendicular to strike.

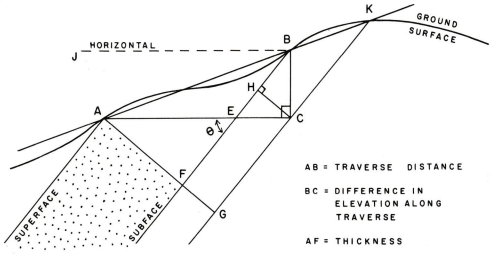

AB = TRAVERSE DISTANCE

BC = DIFFERENCE IN ELEVATION ALONG TRAVERSE

AF = THICKNESS

FIG. 3–4 Thickness calculation where ground slopes opposite to dip direction, with traverse perpendicular to strike.

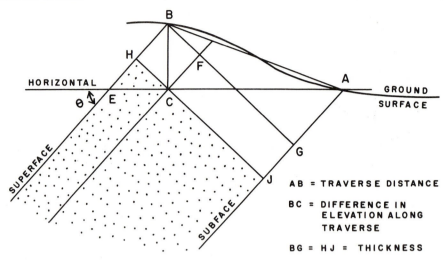

AB = TRAVERSE DISTANCE

BC = DIFFERENCE IN ELEVATION ALONG TRAVERSE

BG = HJ = THICKNESS

perform the trigonometric calculations by slide rule, and then visually compare the computed thickness with the thickness of the outcrop. Gross errors are caught immediately. The geologist leaves the outcrop with the single number he desires—the bed thickness, and he saves the time and inconvenience of drafting a scale drawing.

74. A bluff is formed by horizontal St. Peter Sandstone. A geologist with an eye height of 5 feet 6 inches uses a hand level to measure the vertical distance from the base to the top of the formation. He records 14 eye-heights plus 2 feet. How thick is the St. Peter Sandstone? (Answer: 79 feet)

75. A small butte is capped by an outlier of the White River Group. A geologist with an eye height of 5 feet 8 inches uses a hand level to measure from the base of the White River Group to the top of the butte. What is the thickness of the portion of the White River Group exposed on the butte, if he measures a vertical distance of 8 eye-heights plus 3 feet?

76. A vertical pegmatite dike striking N53°E crops out on a hillside. One contact of the dike is at an altitude of 1,732 feet. The other contact is located 150 feet downslope in a direction S37°E from the first and at altitude 1,682 feet. How thick is the dike? (Answer: 141 feet)

77. A vertical diabase dike striking N27°E crops out on a hillside. A contact on one side of the dike is at an elevation of 763 feet. The other contact is located 123 feet upslope in a direction N63°W from the first and at 797 feet elevation. How thick is the dike?

78. A vertical exposure of the Clinch Sandstone strikes N45°E. A traverse across the outcrop extends for 327 feet N7°E directly down a hill sloping 18°. How thick is the sandstone?

79. A vertical exposure of the Sundance Formation strikes N23°W. A traverse across

the outcrop extends 475 feet N54°W along a hill sloping 13° in the direction of traverse. What is the thickness of the formation?

80. The Frog Mountain Sandstone dips 32°S63°W. A well started at the superface is drilled vertically 49.6 feet deep. What stratigraphic thickness has been penetrated?

81. A well is drilled in an area with a dip 27°S48°E. The Oriskany Sandstone occupies the depth interval 130–217 feet in the vertical well. What is the stratigraphic thickness of the sandstone?

82. The attitude of the Morrison Formation is N30°W 29°NE. The distance across the formation is 438 feet in a traverse measured along a road that slopes 6° in a direction N60°E. How thick is the formation?

83. The attitude of the Athens Shale is N79°E 32°SE. The distance across the formation is 532 feet measured in a traverse along a road that slopes 5° in a direction S11°E. What is the thickness of the Athens Shale?

84. The Midway Formation dips 4°S5°E. The superface and subface are at the same altitude, and the map distance between them is 4,420 feet along a line trending N5°W. How thick is the formation?

85. The Natchez Formation dips 3°S10°W. The top and bottom contacts are at the same elevation, and the width of the outcrop belt is 1,730 feet along a line trending N10°E. How thick is the formation?

86. The attitude of the Lee Formation is N62°E 17°N28°W. The superface and subface are respectively 1,830 and 1,650 feet in elevation. The map distance between these two points along a line extending N28°W is 3,750 feet. Calculate the thickness of the formation.

87. The attitude of the Chugwater Formation is N89°W 13°NE. The superface and subface are respectively 4,215 and 4,025 feet in altitude. The map distance between these two points along a line extending N1°E is 3,200 feet. What is the thickness of the formation?

88. Calculate the thickness of the Clinton Group if it has an attitude of N22°E 28°SE and its thickness is measured by a traverse that is 725 feet in length. The traverse slopes opposite to the dip, and the difference in altitude along the traverse is 238 feet.

89. Calculate the thickness of the Arkansas Novaculite if it has an attitude of N72°W 52°SW and its thickness is measured by a traverse that is 425 feet long. The traverse slopes opposite to the dip, and the difference in altitude along the traverse is 193 feet.

90. The attitude of the Oneota Dolomite is N80°W 2°SW. The elevations of the superface and subface are 853 and 895 feet respectively. The outcrop of the superface is 4,250 feet S10°W of the subface. Calculate the thickness of the dolomite.

91. The attitude of the Wilcox Formation is W 3°S. The altitudes of the superface and subface are 487 and 205 feet respectively. The outcrop of the superface is 9,650 feet due south of that of the subface. Calculate the thickness of the formation.

92. The Gasper Limestone is measured along a road trending N63°E. The attitude of the formation is N27°W 6°SW. The respective altitudes of the superface and sub-

face are 1,427 and 1,346 feet, and the distance between the two points is 1,890 feet along the sloping road. How thick is the limestone?

93. The Fort Payne Chert is measured along a line trending N47°E. The attitude of the chert bedding is N43°W 55°SW. The elevation of the superface is 1,196 feet, and the elevation of the subface is 1,136 feet. The traverse distance between the two points is 268 feet. How thick is the formation?

The General Case

Traverse Not Perpendicular to Strike of Dipping Beds (*Sloping ground surface*) **/** This is the general case of thickness problems, and all of the preceding types are special simplified cases. The formulae are slightly different depending on whether the ground slopes in the same or opposite direction as the dip of the beds.

A. Ground slope and dip are in the same general directions.

The following derivations can be obtained from Figure 3–5.

Thickness = AH = AJ − HJ = AJ − EG.

Thickness = EA (Sin AEJ) − FE (Cos FEG).

Thickness = CA (Cos CAE) (Sin AEJ) − FE (Cos AEJ).

Thickness = AB (Cos BAC) (Cos CAE) (Sin AEJ) − BC (Cos AEJ).

Thickness = (Traverse distance)(Cos traverse slope angle)(Cos angle between traverse and dip directions)(Sin dip angle) − (Traverse distance)(Sin slope angle)(Cos dip angle). (20)

Other formulae that can be derived from this situation are:

Thickness = (Traverse distance) [(Cos traverse slope angle)(Sin angle between traverse and strike directions)(Sin dip angle) − (Sin slope angle)(Cos dip angle)]. (21)

Thickness = (Traverse distance) $\left(\text{Cos arc sin } \dfrac{\text{Difference in altitude along traverse}}{\text{Traverse distance}}\right)$ (Cos angle between dip and traverse directions)(Sin dip angle) − (Difference in altitude along traverse)(Cos dip angle). (22)

Thickness = (Horizontal distance along traverse) (Cos angle between traverse and dip directions)(Sin dip angle) − (Difference in altitude along traverse)(Cos dip angle). (23)

Formula 23 is especially useful with topographic or plane table maps.

B. Ground slope and dip are in opposite general directions.

Figure 3–6 illustrates this situation, and the following formulae can be derived from it.

Thickness = (Traverse distance)(Cos traverse slope angle)(Cos angle between traverse and dip directions)(Sin dip angle) + (Traverse distance)(Sin slope angle)(Cos dip angle). (24)

Thickness = (Traverse distance) [(Cos traverse slope angle) (Sin angle between traverse and strike directions) (Sin dip angle) + (Sin slope angle) (Cos dip angle)]. (25)

Thickness = (Traverse distance) $\left(\text{Cos arc sin} \dfrac{\text{Difference in altitude}}{\text{Traverse distance}}\right)$ (Cos angle between dip and traverse directions) (Sin dip angle) + (Difference in altitude) (Cos dip angle). (26)

Thickness = (Horizontal distance along traverse) (Cos angle between traverse and dip directions) (Sin dip angle) + (Difference in altitude) (Cos dip angle). (27)

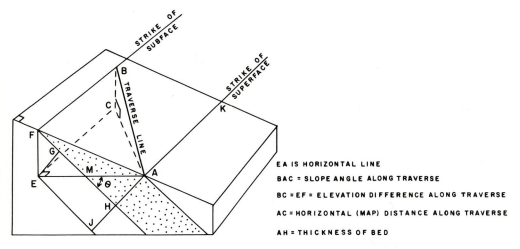

EA IS HORIZONTAL LINE

BAC = SLOPE ANGLE ALONG TRAVERSE

BC = EF = ELEVATION DIFFERENCE ALONG TRAVERSE

AC = HORIZONTAL (MAP) DISTANCE ALONG TRAVERSE

AH = THICKNESS OF BED

FIG. 3–5 Thickness calculations on surfaces inclined *in general direction of dip.*

FIG. 3–6 Thickness calculations on surfaces inclined *in general direction opposite to dip.*

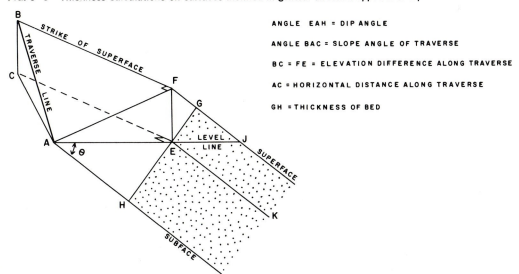

ANGLE EAH = DIP ANGLE

ANGLE BAC = SLOPE ANGLE OF TRAVERSE

BC = FE = ELEVATION DIFFERENCE ALONG TRAVERSE

AC = HORIZONTAL DISTANCE ALONG TRAVERSE

GH = THICKNESS OF BED

The graphic solution for the general problem diagrammed in Figure 3–6 is shown in Figure 3–7. Letter symbols are the same in both illustrations. The three steps in Figure 3–7 are also present in the trigonometric solution. Graphic solutions are fairly simple and are as accurate as most of the initial data obtained for this type of problem. Trigonometric solutions have the advantages of greater accuracy because of elimination of graphic plotting errors, and computations can be made in the field by slide rule. In the latter instance, the trigonometric thickness can be compared visually with the actual outcrop measured, thereby reducing the chances of large mistakes. The initial data should always be recorded regardless of which type of solution is used. If computations are made in the field, one can be certain that sufficient data have been recorded so that a return trip to the outcrop will not be necessary.

FIG. 3–7 Graphic solution to problem of Figure 3–6.

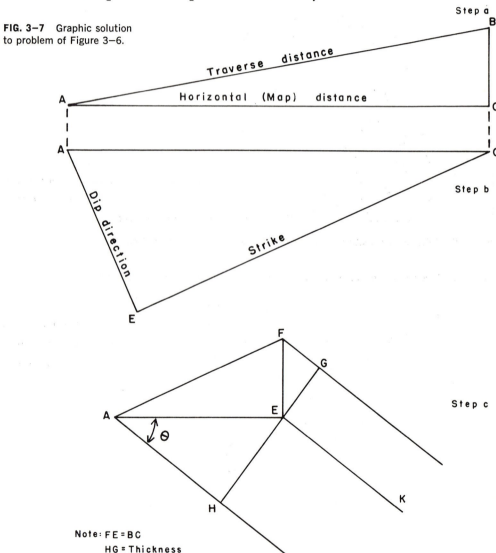

Level Ground Surface

Formula 28 is a special case of Formulae 21 and 25, because no correction need be made for change in elevation along traverse.

Thickness = (Traverse distance)(Sin angle between traverse and strike directions)(Sin dip angle). (28)

Formula 29 is a special case of Formulae 20, 22, 23, 24, 26, and 27.

Thickness = (Traverse distance)(Cos angle between traverse and dip directions)(Sin dip angle). (29)

Beds with Changing Dip

Frequently the attitudes of the superface and subface are not the same, because the dip changes in the interval. If the dip of the superface and subface differ only slightly (not more than 10°) in direction and amount, a rather satisfactory thickness can be calculated from the average angular value of the two dips. An even better solution results from averaging the values of the trigonometric function needed for the computation, using the angular relations of traverse to the superface and to the subface. Of course, both methods assume that the dip changes at a constant rate.

The thickness of a bed with changing dip can be approximated by plotting the outcrops within the bed and their respective dips projected onto a cross section drawn in a plane normal to the average strike. Then carefully sketch the superface and subface in a series of parallel folds. Fairly accurate thickness measurements can be obtained from scale drawings of this sort. This is the only method that will yield an estimate of thickness in a complexly folded section.

Another approach is to divide the traverse across the bed into segments in which the dip change is slight, calculate the thickness of the portion of the bed in each segment, and algebraically add the thickness of the segments to obtain the total thickness of the bed. If the traverse proceeds generally upward stratigraphically, but one segment proceeds downward stratigraphically, the total bed thickness is obtained by subtracting the thickness of the downward-proceeding segment from the total thickness of the upward-proceeding segments.

Nomogram for Computing Thickness

The thickness of a bed can be determined easily by use of a nomogram or alignment diagram, if the traverse is perpendicular to the strike (see Fig. 3–8). If the traverse is across level ground, a ruler is held connecting the mark representing traverse distance (right scale) and the mark corresponding to the dip angle (left scale). The bed thickness is then read at the intersection of the ruler and the middle scale. For example, if the traverse distance is 600 feet and the dip is 30°, the thickness is 295 feet as shown on the diagram. The thickness as calculated trigonometrically is actually 300 feet, and the difference illustrates the accuracy limitation inherent in drafting most alignment diagrams.

The nomogram in Figure 3–8 can also be used if the ground surface slopes and the traverse is perpendicular to strike. If the slope and dip are in the same direction, the dip angle used on the alignment diagram is equal to the true dip angle minus the ground slope angle. If the slope and dip are in opposite directions, the dip angle used on the alignment diagram equals the true dip angle plus the slope angle.

There is also a polyphase alignment diagram for cases in which the traverse is not perpendicular to strike. The construction of such a diagram is described by Palmer (1916) and by Mertie (1922, pp. 39–52). Mertie's diagram is reproduced by Billings (1954, Fig. 350).

FIG. 3–8 Nomogram or alignment diagram for use in thickness determinations. Dashed line indicates situation where the traverse distance is 600 feet and the dip angle is 30°. The resultant thickness is 295 feet. (*After H. S. Palmer, 1916.*)

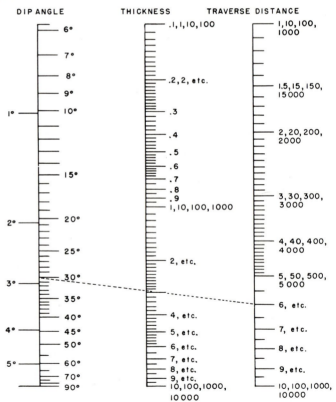

Exercises /

94. A level exposure of Chattanooga Shale extends for 200 feet perpendicular to strike. The dip is 19°N32°W. How thick is the shale as calculated by the alignment diagram? Check by the trigonometric or graphic method.

95. The superface and subface of the Marcellus Shale are at the same elevation and

are separated by a horizontal distance of 632 feet along a line trending perpendicular to strike. The dip is 49°N45°W. How thick is the shale as calculated by the alignment diagram? Check by the trigonometric or graphic method.

96. The Johns Valley Shale dips 39°S32°E. A traverse along a road sloping 3°N32°W extends 223 feet from superface to subface of the shale. How thick is the exposure? Use the nomogram, and then check by another method.

97. The Dakota Sandstone has an attitude of N17°W 32°NE. A traverse across the formation is 752 feet long and extends along a road sloping 5°N73°E. Determine the formation thickness using the alignment diagram, and check by the graphic or trigonometric method.

98. The Orinda Formation has an attitude of N47°W 29°SW. A traverse across the formation is 1,320 feet long and extends along a road sloping 4°S43°W. Calculate the formation thickness using the alignment diagram, and check by the graphic or trigonometric method.

99. The Oriskany Sandstone dips 15°S31°E. A traverse along a road sloping 5°N31°W is 173 feet long from top to bottom of the sandstone. How thick is the exposure? Use the nomogram, and check by another method.

100. The attitude of the Bedford Limestone is N69°E 4°NW. A traverse across the exposure extends for 1,340 feet in the direction N45°W. The superface and subface attitudes are 893 and 837 feet respectively. How thick is the limestone?

101. The Hartselle Sandstone dips 17°S57°W. A traverse along a road cut slopes 8°S23°W and is 124 feet long. Determine the thickness of the sandstone.

102. An outcrop of Dakota Sandstone extends for 283 feet along the water's edge of a lake, the shore of which trends N37°W. The attitude of the superface is 23°N3°W, and the subface attitude is 29°N4°E. How thick is the sandstone?

103. The attitude of the Allegheny Series is N32°E 7°NW. A traverse across the outcrop extends for a distance of 1,435 feet in a direction N45°W. Elevations of the superface and subface are 982 and 957 feet respectively. What is the stratigraphic thickness of the Allegheny Series?

104. An outcrop of the Helderberg Group extends for 657 feet along the shore of a reservoir which trends N47°W. The attitude of the superface is 37°N13°W, and the subface attitude is 33°N19°W. How thick is the Helderberg?

105. The Huntersville Chert dips 67°N42°W. A traverse along a stream bank slopes 13°N23°W and is 73.2 feet long. Determine the thickness of the chert.

106. An irregular traverse extends across the Parkwood Formation. Calculate the thickness of the formation based on data taken at stations between the superface and subface.

	SUBFACE	POINT A	POINT B	POINT C	SUPERFACE
ATTITUDE OF BED	N21°E 42°SE	37°S65°E	35°S70°E	29°S67°E	25°S72°E
ALTITUDE (FEET)	786	832	797	843	916
TRAVERSE DISTANCE TO NEXT POINT (FEET)	432	237	177	519	
DIRECTION TO NEXT POINT	S87°E	N24°E	S75°E	S42°E	

107. The following traverse was taken across the Phosphoria Formation. Calculate its thickness.

	SUBFACE	POINT A	POINT B	POINT C	SUPERFACE
ATTITUDE OF BED	17°S78°W	12°S73°W	5°S80°W	Horizontal	3°N75°E
ALTITUDE (FEET)	6725	6762	6791	7013	7047
TRAVERSE DISTANCE TO NEXT POINT (FEET)	217	179	445	199	
DIRECTION TO NEXT POINT	S32°W	S57°W	S63°W	S45°W	

108. An irregular traverse extends from the bottom to top of the Mauch Chunk Group. Calculate its thickness from the following data taken at stations between the superface and subface.

	SUBFACE	POINT A	POINT B	POINT C	SUPERFACE
ATTITUDE OF BED	N43°E 25°SE	17°S40°E	8°S44°E	2°S46°E	11°S40°E
ALTITUDE (FEET)	1690	1690	1903	1876	1876
TRAVERSE DISTANCE TO NEXT POINT (FEET)	498	735	67	1170	
DIRECTION TO NEXT POINT	S27°E	S13°E	S45°E	S22°E	

109. The following series of traverse was taken across an exposure of Lockport Dolomite. Calculate the thickness of the dolomite.

	SUBFACE	POINT A	POINT B	POINT C	SUPERFACE
ATTITUDE OF BED	10°S4°E	13°S10°E	7°S2°E	3°S4°W	Horizontal
ALTITUDE (FEET)	358	370	370	302	338
TRAVERSE DISTANCE TO NEXT POINT (FEET)	257	488	365	842	
DIRECTION TO NEXT POINT	S30°E	S6°E	S43°W	S13°E	

110. The subface of a formation crops out at point C in Figure 3–9, where the attitude of the strata is N27°E 12°NW. The superface crops out at point B where the beds dip 14°S65°W. How thick is the formation?

111. The subface of a formation crops out at point B in Figure 3–9, where the beds dip 23°N19°E. The attitude of the superface at point A is N69°W 25°NE. How thick is the formation?

112. The subface of a formation crops out at point A in Figure 3–9, where the attitude of the strata is 33°S16°E. The superface crops out at point B where the beds dip 29°S20°E. How thick is the formation?

113. The subface of a formation crops out at point C in Figure 3–9, where the beds dip 12°N5°W. The attitude of the superface at point A is 9°N1°E. How thick is the formation?

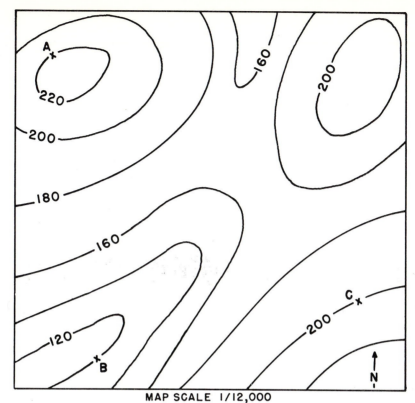

FIG. 3–9
Map for Exercises 110–113.

MAP SCALE 1/12,000

References

Billings, M. P., 1954, Structural geology: New York, Prentice-Hall, Inc., 2nd ed., 514 p.

Hansen, W. R., 1960, Improved Jacobs Staff for measuring inclined stratigraphic intervals: Am. Assoc. Petroleum Geologists Bull., v. 44, p. 252–255.

Hewett, D. F., 1920, Measurements of folded beds: Econ. Geology, v. 15, p. 367–385.

Hobson, G. D., 1942, Calculating the true thickness of a folded bed: Am. Assoc. Petroleum Geologists Bull., v. 26, p. 1827–1831.

Ickes, E. L., 1925, The determination of formation thickness by the method of graphical integration: Am. Assoc. Petroleum Geologists Bull., v. 9, p. 451–463.

Kottlowski, F. E., 1965, Measuring stratigraphic sections: New York, Holt, Rinehart, and Winston, 253 p.

Lahee, F. H., 1961, Field geology: New York, McGraw-Hill Book Co., 6th ed., 926 p.

LeRoy, L. W., and Low, J. W., 1954, Graphic problems in petroleum geology: New York, Harper and Bros., p. 140–143, 158–161. (Extension of Busk method of drawing folds to three dimensions, and using it to calculate stratigraphic thickness.)

Mertie, J. B., Jr., 1922, Graphic and mechanical computation of thickness of stratum and distance to a stratum: U.S. Geol. Sur., Prof. Paper 129-C, 52 p.

———, 1940, Stratigraphic measurements in parallel folds: Geol. Soc. America Bull., v. 51, p. 1113–1122.

———, 1944, Calculation of thickness in parallel folds: Am. Assoc. Petroleum Geologists Bull., v. 28, p. 1376–1385.

Miller, F. S., 1944, Graphs for obtaining true thickness of a vein or bed: Amer. Inst. Min. Eng. Contrib. No. 136, 6 p.

Palmer, H. S., 1916, Nomographic solution of certain stratigraphic measurements: Econ. Geology, v. 11, p. 14–29.

———, 1919, New graphic method for determining the depth and thickness of strata and the projection of dip: U.S. Geol. Sur., Prof. Paper 120, p. 123–128.

Robinson, G. D., 1959, Measuring dipping beds: Geotimes, v. 4, no. 1, p. 8–9, 24–25, 27.

Secrist, M. H., 1941, Computing stratigraphic thickness: Amer. Jour. Sci., v. 239, p. 417–420.

Threet, R. L., 1961, A simplified slide rule for stratigraphic section measurement: Jour. Geological Education, v. 9, p. 70–73.

CHAPTER 4

Depth to Dipping Strata

GEOLOGISTS ENGAGED IN SUBSURFACE WORK of any kind must be able to calculate the depth to a formation from one place on the ground surface, using information obtained from outcrops of that formation at one or more other places. Applications of this include the tracing of a petroleum reservoir rock to estimate the depth at which it may be intercepted in a well, the recommendation of depth of core drilling to find the character and extent of a metalliferous vein, and the determination of cost of drilling a water well to a certain horizon. In an area where some subsurface work has already been done, the outcrop information can be supplemented by subsurface data.

Depth is the vertical distance from the ground surface to the horizon in question. Graphic, trigonometric, or alignment-diagram solutions can be used to obtain the depth to a stratum or other geologic surface (such as a fault, vein, or the contact between country rock and a stock). Trigonometric methods are emphasized in the following examples, but graphic solutions may be obtained from scale drawings in which the geometric relationships are plotted and the depth is measured directly.

Depth to Planar Units

Horizontal Strata / The depth to a horizontal stratum is, of course, the difference in altitude between the stratum and the ground surface.

Strata with Uniform Dip /

A. Traverse perpendicular to strike.

This special case of the depth problem is encountered in inclined beds or in folded mountain regions.

1. Horizontal ground surface.

The following derivation can be obtained from Figure 4–1.

BC = Depth to stratum.

$\frac{BC}{AB}$ = Tan BAC.

BC = AB (Tan BAC).

Depth = (Horizontal distance)(Tan dip angle). (30)

FIG. 4–1 Depth determination to dipping bed; traverse at right angles to strike, ground surface level.

FIG. 4–2 Depth determination to bed dipping opposite to surface slope; traverse at right angles to strike.

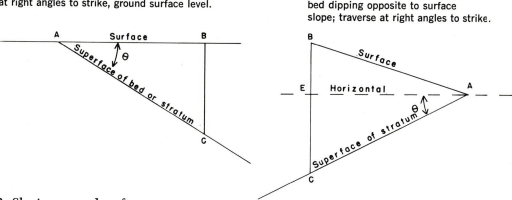

2. Sloping ground surface.

a. Slope and dip in opposite directions.

Figure 4–2 is the reference for the following derivation.

BC = Depth to stratum.

BC = CE + BE.

BC = EA (Tan EAC) + BA (Sin BAE).

BC = BA (Cos BAE) (Tan EAC) + BA (Sin BAE).

BC = BA [(Cos BAE) (Tan EAC) + (Sin BAE)].

Depth = (Traverse distance)[(Cos slope angle)(Tan dip angle) + (Sin slope angle)]. (31)

Often the difference in altitude along a known slope is easier to obtain than a slope angle. In that case:

$$\textbf{Depth = (Traverse distance)}\left(\textbf{Cos arc sin} \frac{\textbf{Difference in altitude along traverse}}{\textbf{Traverse distance}}\right)$$ (32)
(Tan dip angle) + (Difference in altitude along traverse).

It is often desirable to calculate the depth to a formation using a topographic map to obtain the horizontal distance and difference in altitude along the traverse. In this situation:

Depth = (Horizontal traverse distance)(Tan dip angle) + (Difference in altitude along traverse). (33)

b. Slope and dip in the same direction.

Formulae for this situation can be derived from Figure 4–3 in a manner similar to those developed from Figure 4–2 where the dip and slope are in opposite directions.

Depth = (Traverse distance)[(Cos slope angle)(Tan dip angle) − (Sin slope angle)]. (34)

$$\text{Depth} = (\text{Traverse distance}) \left(\text{Cos arc sin} \frac{\text{Difference in altitude along traverse}}{\text{Traverse distance}} \right)$$ (35)
(Tan dip angle) − (Difference in altitude along traverse).

Depth = (Horizontal traverse distance)(Tan dip angle) − (Difference in altitude along traverse). (36)

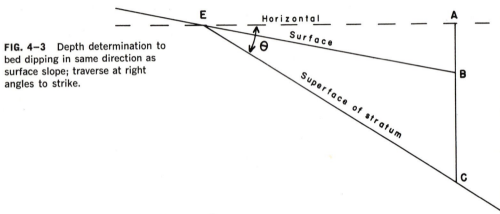

FIG. 4–3 Depth determination to bed dipping in same direction as surface slope; traverse at right angles to strike.

B. Traverse not perpendicular to strike.

This general type of depth problem is probably the most common kind of depth computation.

1. Level ground surface.

Figure 4–4 can be used to derive the following formula.

Depth = (Horizontal traverse distance)(Cos angle between traverse and dip directions)(Tan dip angle). (37)

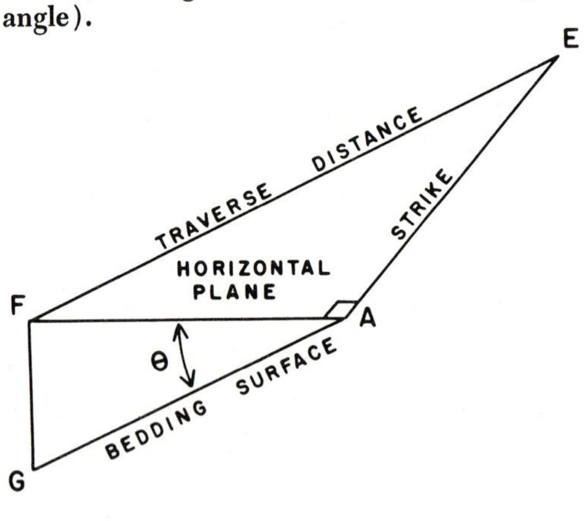

FIG. 4–4 Depth determination to dipping bed; traverse oblique to strike, ground surface level.

2. Sloping ground surface.
a. Dip and traverse slope in opposite general directions.

The following derivation can be obtained from Figure 4–5.

Depth = BJ, BE = CF, EJ = FG.

BJ = EJ + BE.

BJ = FG + BE.

BJ = AF (Tan FAG) + BE.

BJ = AE (Cos EAF) (Tan FAG) + AB (Sin BAE).

BJ = AB (Cos BAE) (Cos EAF) (Tan FAG) + AB (Sin BAE).

BJ = AB [(Cos BAE) (Cos EAF) (Tan FAG) + (Sin BAE)].

Depth = (Traverse distance) [(Cos traverse slope and angle) (Cos angle between traverse and dip directions) (Tan dip angle) + (Sin traverse slope angle)]. (38)

FIG. 4–5 Depth determination to dipping bed; traverse not perpendicular to strike, dip in opposite general direction from surface (traverse) slope.

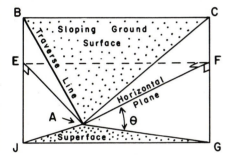

ACFG is true dip plane.

ABEJ is vertical plane containing traverse line.

FAG = Dip angle.

EAF = Angle between directions of traverse and true dip, measured in horizontal plane.

Another variation of this formula is

Depth = (Traverse distance) [(Cos traverse slope angle) (Sin angle between traverse and strike directions) (Tan dip angle) + (Sin traverse slope angle)]. (39)

If the traverse length and the difference in altitude along the traverse are determined, the following formula can be used:

$$\text{Depth} = \text{(Traverse distance)} \left(\text{Cos arc sin} \frac{\text{Difference in altitude along traverse}}{\text{Traverse distance}} \right) \quad (40)$$
(Cos angle between traverse and dip directions) (Tan dip angle) + (Difference in altitude along traverse).

The following formula can be used if the difference in altitude and the horizontal distance along the traverse are determined from a topographic map:

Depth = (Horizontal traverse distance) (Cos angle between traverse and dip directions) (Tan dip angle) + (Difference in elevation along traverse). (41)

b. Dip and traverse slope in same general directions.

Formulae for this situation are obtained in a manner similar to derivations for the case in which the traverse slope and dip are in opposite general directions.

Depth = (Traverse distance) [(Cos traverse slope angle)(Cos angle between traverse and dip directions)(Tan dip angle) − (Sin traverse slope angle)]. (42)

Depth = (Traverse distance) [(Cos traverse slope angle)(Sin angle between traverse and strike directions)(Tan dip angle) − (Sin traverse slope angle)]. (43)

Depth = (Traverse distance) $\left(\text{Cos arc sin} \dfrac{\text{Difference in altitude along traverse}}{\text{Traverse distance}}\right)$ (Cos angle between traverse and dip directions)(Tan dip angle) − (Difference in altitude along traverse). (44)

Depth = (Horizontal traverse distance)(Cos angle between traverse and dip directions)(Tan dip angle) − (Difference in altitude along traverse). (45)

Graphical methods can be used in place of the trigonometric solutions for each of the cases described above. For traverse at right angles to strike, prepare drawings similar to Figures 4–1 to 4–3, using the proper angles and scaled distances.

Figure 4–6 shows the graphic solution to the more complex problem of Figure 4–5. Lettered points are the same for both figures. In Step A the sloping traverse distance is projected into the horizontal plane to yield map distance along traverse. In Step B this map distance is projected along strike to yield distance perpendicular to strike. Finally in Step C a scaled cross section is drawn at right angles to strike, and the depth to the stratum is measured as CG.

FIG. 4–6 Graphic solution for problem of Figure 4–5. General case of depth determination to dipping bed; traverse not perpendicular to strike, ground surface slopes in opposite general direction from dip.

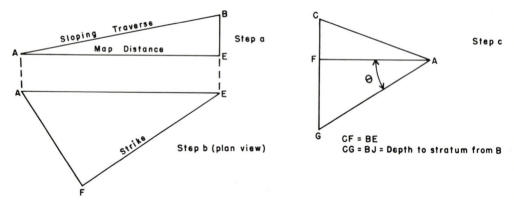

Strata with Changing Dip

It is difficult to obtain accurate estimates of the depth to a horizon if the dip changes laterally. If an estimated value is known for the dip of the horizon at the point of its intersection by the boring, and if this estimated dip does not differ significantly from the outcrop dip, the depth can be approximated by using the average angular value of the two dips. The method of using average angular values may be used if the difference of the values is less than about 10° for dip directions and less than about 10° for dip amounts smaller than 60°. The error increases rapidly for dip amounts greater than 60°. A more accurate approximation can be obtained by using an average value of each necessary trigonometric function based on the two dips. Of course, both of these methods assume that the rate of change of dip is uniform.

Another satisfactory procedure is to construct an accurate structure section with the same horizontal and vertical scale and measure directly the desired depth in this structure section drawn perpendicular to the average strike. This is the easiest means of estimating depth to a specific horizon in an area with complex structure.

Nomogram for Calculating Depth to Strata

The alignment diagram in Figure 4–7 can be used only if the traverse distance is measured perpendicular to the strike.

The simplest use is for horizontal ground surfaces. Place a ruler on the diagram connecting the points indicating the proper traverse distance in feet and the dip in degrees; the depth to the horizon is read at the intersection of the ruler and the middle scale of the alignment diagram. For instance, if the traverse distance is 300 feet and the dip is 10°, the depth to the stratum is 52 feet.

The diagram can also be used when the ground surface is sloping perpendicularly to the strike. If the dip and slope are in opposite directions, the slope angle is added to the dip angle, and this value is used on the dip degrees scale of the nomogram. If the dip and slope are in the same direction, the value of the dip angle minus the slope angle is used. The distance used on the diagram is the actual length of the sloping traverse, not the map or horizontal length. This is the *only* situation in depth problems where additions and subtractions of slope angles are performed.

The sloping ground surface situation can also be solved with the alignment diagram by using the horizontal distance along the sloping traverse and the true dip angle. The difference in altitude along the traverse is then added to or subtracted from the alignment diagram depth, depending on whether the dip and slope are in opposite or the same directions, respectively.

It is impractical to use the alignment diagram in Figure 4–7 for a traverse oblique to strike, because all distances and slope angles would have to be projected onto a plane normal to the strike. Special alignment diagrams have been developed by Mertie (1922) for determining depths from traverse distances measured oblique to the strike. Such a diagram has been reproduced by Billings (1954, Fig. 351).

Bed Thickness in a Well

The distance through a dipping stratum penetrated by a vertical well exceeds the true thickness of the bed. The apparent thickness of an inclined stratum is the vertical distance from the superface to the subface. The following formulae are valid if the well is vertical.

Well thickness = (Stratigraphic thickness) (Sec dip angle). (46)

Stratigraphic thickness = (Well thickness) (Cos dip angle). (47)

Distance along Inclined Hole

Thus far depth has been considered simply the distance along a vertical line from the ground surface to the intersection with some other reference surface. Two deviations from this simple case are sometimes encountered. If an inclined well is drilled to an approximate horizontal surface (such as the level summit of a dome containing natural gas), the apparent depth (well distance) is somewhat greater than in a vertical hole (see Fig. 4–8). The vertical distance to the bedding surface FC is the length of line AC. The inclined distance is longer.

$$AC = AF(\text{Cos FAC}).$$

$$AF = \frac{AC}{\text{Cos FAC}}.$$

$$AF = AC(\text{Sec FAC}).$$

Inclined distance = (Vertical distance) (Sec angle between bore and vertical). (48)

Distance through a horizontal stratum (apparent thickness) is shown by

Apparent thickness = (True thickness) (Sec angle between bore and vertical). (49)

Conversely, for drill holes penetrating flat strata

Vertical depth = (Apparent depth) (Cos angle between bore and vertical). (50)

Formula 50 is commonly used to determine the true vertical depth in petroleum wells whose bore has wandered off a truly vertical path.

The second type of deviation from a vertical hole is the general case of distance along an inclined hole (Fig. 4–9). In the general case the inclined hole is not normal to the bedding plane. *Suppose the top of a hematite bed crops out at point A in Figure 4–9, with an elevation of 980 feet. The attitude of the hematite bed is N50°W 17°NE. A core is bored at point B located 200 feet map distance N60°E of point A. The ground elevation at the top of the well is 1,050 feet. The core slants in a direction N70°W from the collar of the hole, with a hole inclination of 30° from the vertical. At what distance along the inclined hole will the hematite bed be encountered?* A map view of the situation is shown at the top of Figure 4–9. The graphical solution is obtained from a scale drawing along a section trending N70°W (the direction of plunge of the core). The apparent dip of the hematite in that plane is 6°S70°E according to the procedure outlined in Chapter 1 (Formula 1 or Fig. 1–8). A scaled cross section in a vertical plane trending N70°W is shown in the second step of Figure 4–9. Line BF is vertical. Distance EC equals CB from Step 1 of Figure 4–9. Angle HBF is 30°. The length of the inclined hole BH is 139 feet.

FIG. 4–7 Nomogram or alignment diagram for use in depth determinations. Dashed line indicates situation where the traverse distance is 300 feet and the dip angle is 10°. (*After H. S. Palmer*, 1918.)

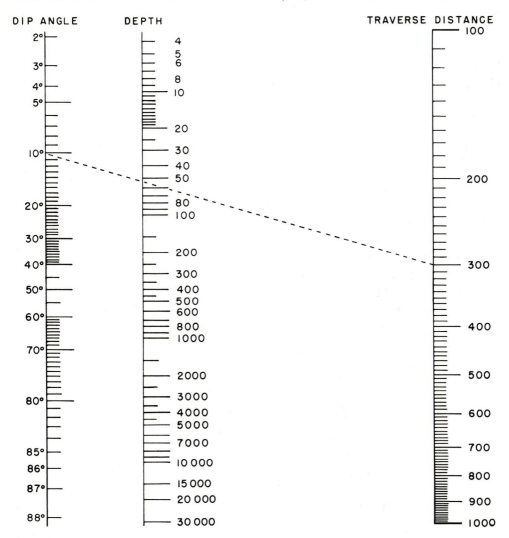

FIG. 4–8 Inclined bore intersecting horizontal strata.

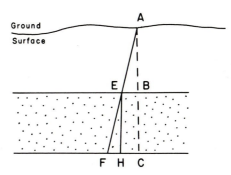

It is possible to develop a totally trigonometric solution for the situation in Figure 4–9 by using the law of sines in both the map view and the cross-section view of the figure.

Exercises /

114. At an outcrop on level ground, the superface of a limestone has an attitude of N77°E 21°NW. A well drilled 430 feet N13°W of the outcrop must attain what depth in order to reach the limestone? Solve trigonometrically, graphically, and by using the alignment diagram.

115. The superface of a level outcrop of sandstone has an attitude of N37°W 33°NE. A well drilled 715 feet N53°E of the superface outcrop will penetrate the super-face at what depth? Solve trigonometrically, graphically, and by using the alignment diagram.

116. The dip of the superface of the Clinton iron ore is 28°S72°E in an outcrop at the bottom of a hillside sloping 8° in the direction of dip. A vertical drill core sample is to be taken at a point 290 feet slope distance from the superface outcrop. At what depth would the top of the ore be expected? Solve trigonometrically, graphically, and by using the nomogram. (Answer: 112 feet)

117. The upper surface of a quartz vein at an altitude of 3,741 feet dips 68°N45°W on a mountainside that slopes N45°E. At what depth should the vein be reached in a vertical boring located 668 feet directly up the slope from the outcrop and at an altitude of 3,917 feet? Solve graphically and trigonometrically.

118. The Pittsburgh coal dips 72 feet per mile in a direction N65°W. If the top of the coal crops out at 957 feet elevation, to what depth must a shaft be dug in order to reach the coal at a point 12,500 feet N65°W of the outcrop and with a ground surface elevation of 1,043 feet? Solve graphically and trigonometrically.

119. The top of a sandstone impregnated with asphalt crops out at an altitude of 837 feet, and it dips 42°S28°E beneath an impermeable shale. A wildcat well with surface altitude of 1,140 feet is located 910 feet ground distance along a uniform slope S28°E from the outcrop. At what depth should the sandstone be reached? Solve graphically and trigonometrically.

120. The dip of the superface of a sphalerite-bearing limestone is 25°S42°W. A drill-core sample is to be taken at a point 380 feet slope distance down the hillside from the superface outcrop. The hillside slopes 6° in the direction of dip. At what depth should the top of the ore-bearing limestone be expected? Solve trigonometrically and graphically, and by using the alignment diagram.

121. The upper surface of a quartz vein at an altitude of 2,372 feet dips 65°N32°E on a mountainside that slopes S32°W. At what depth should the vein be reached in a boring located at an altitude of 2,527 feet and 970 feet directly up the slope from the outcrop? Solve graphically and trigonometrically.

122. The contact of the Wilcox and Midway formations crops out at an altitude of 156 feet, and the horizon dips 5°S7°E. The location is plotted on a topographic map.

FIG. 4—9 General case of distance in inclined hole. Hole is inclined in different direction from dip of bed. In Step 1, use scale drawing to determine distance BC and then ascertain apparent dip in BC direction. In Step 2, use scale drawing to determine inclined hole distance BH.

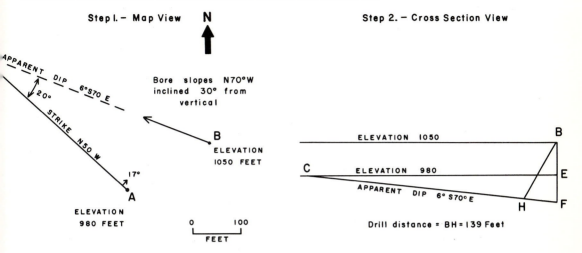

A well site is located according to the map at a point 900 feet S7°E from the exposure and at an altitude of 273 feet. At what depth should the well penetrate the horizon? Solve graphically and trigonometrically, and check with the nomogram method.

123. The top of a sandstone impregnated with asphalt crops out at an altitude of 1,232 feet, and it dips 35°E beneath an impermeable shale. A wildcat well with a surface altitude of 1,187 feet is located one-eighth mile ground distance due east of the outcrop. The ground slopes uniformly between the outcrop and well. At what depth should the sandstone be reached? Solve graphically and trigonometrically.

124. An outcrop of the superface of the Cayugan salt is located at an elevation of about 540 feet and has a dip of 3°S5°W. A vertical shaft is to be dug to the salt at a distance of 3.17 miles S32°E from the superface outcrop point. The shaft starts at an elevation of 725 feet. At what depth should the shaft reach the salt? Use graphical and trigonometric solutions.

125. An outcrop of the Pine Mountain fault dips 52°S20°E and occurs at an elevation of 1,245 feet. At a point with elevation 1,873 feet and located 1.73 miles S18°W from the outcrop point, it is proposed to drill a test well to determine if the dip of the fault flattens to the southeast of the surface exposure of the fault. Assuming that the dip either remains constant or flattens, what is the maximum depth that would have to be drilled before reaching the fault? Solve graphically and trigonometrically.

126. An outcrop of the superface of the Pittsburgh coal at an altitude of 1,452 feet dips 3°N30°W. At a ground distance of 3,250 feet N77°W of the outcrop along a uniform slope and at an altitude of 1,739 feet, find the distance a vertical shaft must be sunk to reach the coal. Use graphic and trigonometric methods.

127. An outcrop of the superface of the Dakota Sandstone dips 17°S37°E. A water well is drilled 2,000 feet slope distance from the exposure at the base of a slope inclined 6°S2°E. At what depth would the sandstone be reached? Use graphic and trigonometric methods.

128. The superface of a uraniferous sandstone dips 18°N30°W at point C on the map in Figure 4–10. At what depth should this sandstone be reached in a well drilled at point B? Use graphic and trigonometric methods.

129. The superface of the Clinton iron ore dips 9°N21°E at point B of Figure 4–10. How deep would a vertical shaft at point A have to be in order to reach the ore? Use graphic and trigonometric methods.

130. The superface of a coal bed dips 7°S83°E at point A of Figure 4–10. How deep would a vertical shaft have to be at point C in order to reach the coal?

131. A permeable sandstone dips 42°S89°W from point B of Figure 4–10. How deep would a water well on the hill at point A have to be in order to reach the horizon of the sandstone?

132. A vertical core through a bed of hematite is 17.8 feet long. The dip of the strata is 78°N45°E. How thick is the iron-bearing bed?

133. A well is drilled at point A in Figure 4–10. Steel line measurement indicates the hole is 7950 feet deep. If the hole is inclined an average of 2° from the vertical, what is the elevation of the bottom of the hole?

134. A core hole sloping 30° from the horizontal penetrates 158 feet into a hillside. The elevation of the collar of the hole is 976 feet. What is the elevation of the bottom of the hole?

135. A high-magnesium limestone stratum dips 25°N2°W and crops out at an elevation of 480 feet. A test hole is drilled starting 350 feet map distance N25°E from the outcrop and with a collar elevation of 530 feet. The hole has a slope 30° from the vertical in a direction S53°W. At what drill-hole distance should the stratum be encountered?

136. A copper-bearing conglomerate dips 13°S20°E at an outcrop 750 feet in elevation. At a locality 505 feet slope distance in a direction S32°W from the outcrop a core is drilled starting at an elevation of 920 feet. The core plunges 60°due north. At what drilling distance would you expect to encounter the conglomerate?

FIG. 4–10 Map for Exercises 128–131 and 133.

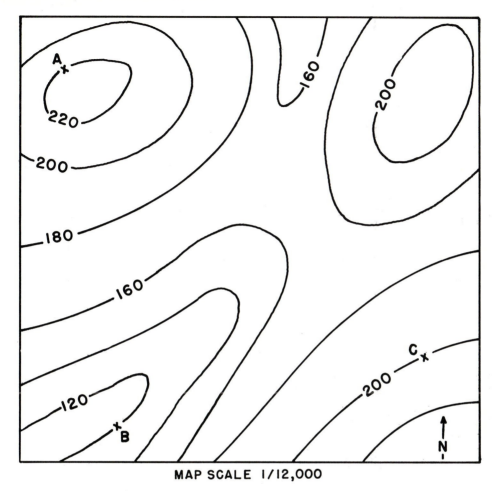

MAP SCALE 1/12,000

References

Billings, M. P., 1954, Structural geology: New York, Prentice-Hall, Inc., 2nd ed., 514 p.

Mertie, J. B., Jr., 1922, Graphic and mechanical computation of thickness of stratum and distance to a stratum: U.S. Geol. Sur., Prof. Paper 129-C, 52 p.

Palmer, H. S., 1919, New graphic method for determining the depth and thickness of strata and the projection of dip: U.S. Geol. Sur., Prof. Paper 120, p. 123–128.

CHAPTER 5

Outcrop Patterns and Three-point Problems

THE *outcrop* OF A FAULT, DIKE, OR BED is the line of intersection of that horizon with the earth's surface. Geologists must frequently plot on a map the outcrop patterns of strata, intrusions, or faults. The problem of outcrop delineation is a study of the locus of intersection of two surfaces in space. If one of these surfaces (the bedding surface) is a plane, the trace on the other surface (the topographic surface) can be determined by simple graphical methods.

Rules of the V's

To work with outcropping strata, it is imperative to understand the relation of the deflection of formation contact lines to dip and topography. These relationships were studied briefly in elementary geology, and are commonly referred to as the "rules (or laws) of the V's." They are listed below and are illustrated in Figure 5–1, maps *A-E*.

A. The contact lines of horizontal beds exactly parallel the topographic contours, and consequently the V's point up the valleys as in Figure 5–1, map *A*.
B. The contact lines of beds regularly dipping at 90° (vertical) will cross a valley with no deflection, as in Figure 5–1, map *B*.
C. If the beds dip opposite to the slope of the valley, the contact lines make a bend pointing up the valley but do not parallel contours. The V's or bends of the contact lines are more blunt than the V's of the contours, as shown in Figure 5–1, map *C*.

FIG. 5–1 Contoured geologic maps illustrating the "Rules of the V's."

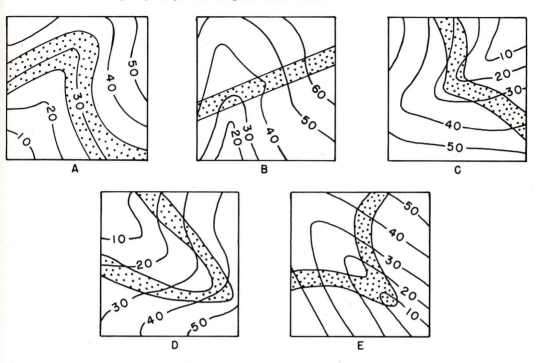

A   B   C

D   E

D. If the beds dip down a valley at a smaller angle than the slope of the stream gradient, the V's of the contact lines point upstream and do not parallel contours. The V's of the contact lines are less blunt than the V's of the contours (Figure 5–1, map *D*).

E. If the beds dip down a valley at a larger angle than the slope of the stream gradient, the V's of the contact lines point downstream, as in Figure 5–1, map *E*.

All five of the rules of the V's can be summarized in a very simple manner (Screven, 1963, p. 98):

The outcrop V points in the direction in which the formation lies underneath the stream.

The strike of a bed whose outcrop crosses a valley can be determined from a combined topographic and geologic map by drawing a line between two points of equal altitude on a formation contact line. The rules of the V's indicate the general dip direction, which is, of course, 90° from the strike. The dip amount can be computed by finding the change in altitude of the formation contact surface in a horizontal distance measured at right angles to the strike. The following formula is pertinent:

$$\text{Tan dip angle} = \frac{\text{Change in altitude of formation contact surface}}{\text{Horizontal distance measured perpendicular to strike}}. \qquad (51)$$

This procedure assumes that the dip is uniform. In areas of changing dip the method yields approximate average dip values.

Three-point Problem

A method for determining dip from the outcrop of three points (with different elevations) on the formation contact line is shown in Figure 5–2. The three points can lie along a continuously exposed outcrop, or they may represent three isolated exposures of the formation boundary. In fact, these may even be three well sites plotted on a map with indication of the elevation of the formation contacts in each well.

Connect the three points to form a triangle (ABC in Figure 5–2) lying in the formation contact plane. The strike is the direction of a level line (CE) in this plane. The length of BE (and consequently the position of point E and the strike direction) can be determined from the following formula:

$$\frac{BE}{AB} = \frac{\text{Difference in altitude between B and C}}{\text{Difference in altitude between A and B}}. \qquad (52)$$

The dip direction is determined by a line (FB) passing through point B and perpendicular to the strike (CE). The dip angle can be calculated from the formula:

$$\text{Tan dip angle} = \frac{\text{Difference in altitude between F and B}}{\text{Horizontal distance between F and B}}. \qquad (53)$$

It should be noted that the altitude of point F is that of the line CE and the point C (500 feet) because point F is located in space above the ground surface.

The three-point method can be used to calculate the dip of any plane in geology: a fault, dike, sill, vein, or joint, as well as a key stratigraphic horizon like a formation contact surface.

Figure 5–3 should be studied along with Figure 5–2 in order to prove Formulae 52 and 53. In Figure 5–3:

Plane A'Bba' is the plane of the formation.

Plane ABba is a horizontal plane at 200 feet altitude.

Triangle ABC and lines CE and BF represent map distance measured on Figure 5–2.

Vertical plane a'ba is parallel to dip direction.

GIVEN:

Elevation of points A', B, and C' and the shape of triangle ABC (angles determined from map view).

TO PROVE:

(1) $\dfrac{BE}{AB} = \dfrac{\text{Difference in altitude between B and C}}{\text{Difference in altitude between A and B}} = \dfrac{C'C}{A'A}.$

(2) $\text{Tan dip angle} = \dfrac{\text{Difference in altitude between F and B}}{\text{Horizontal distance between F and B}}.$

PROOF (1):

$\dfrac{BG}{BA'} = \dfrac{BF'}{BE'}$; $\dfrac{ba'}{bC'} = \dfrac{ba}{bC} = \dfrac{AB}{BE}.$ Corresponding parts of similar triangles.

$\therefore \dfrac{C'C}{a'a} = \dfrac{C'b}{a'b} = \dfrac{BE}{AB},$

FIG. 5–2 Method of determining dip of contact surface knowing positions and elevations of three of its outcrops.

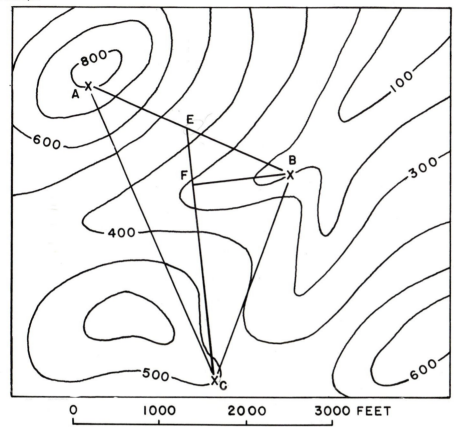

FIG. 5–3 Reconstruction of relationships in Figure 5–2, used for deriving Formulae 52 and 53.

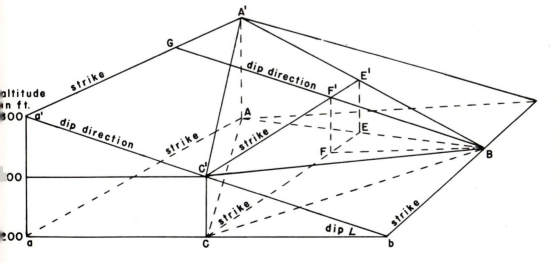

or

$$\frac{\text{Difference in altitude between lowest and intermediate points}}{\text{Difference in altitude between lowest and highest points}} = \frac{BE}{AB}. \qquad (52)$$

PROOF (2):

BF = Cb. By construction.

C'C = Difference in altitude between F and B.

$$\text{Tan dip angle} = \frac{C'C}{bC} = \frac{\text{Difference in altitude between F and B}}{\text{Horizontal distance F to B}}. \qquad (53)$$

Plotting Outcrop Patterns

In a region with constant (homoclinal) dip the outcrop trace of formation contact lines can be determined with great accuracy. In situations where the dip changes slightly in a known manner, but the formation contacts are concealed between certain outcrops, the expected position of the outcrop may be inferred with considerable reliability.

Homoclinal Dips / If the dip of a formation is constant throughout an area, the outcrop trace of the formation contacts can be quickly plotted, provided that there are outcrops of the superface or subface somewhere in the area or that they are located in space from well data. The magnitude and direction of the dip can be measured directly with a Brunton compass, or it can be determined from a three-point solution of three outcrops of the same key horizon.

Figure 5–4 illustrates the method of plotting the outcrop of a formation contact line if the dip and position of an exposure are known. In this example the dip of the formation contact line at point A is 25° due east. Draw a line (BAC) through point A and in the strike direction (due north). The altitude of the formation contact surface is 600 feet all along line BAC. A cross-section grid should be prepared with vertical and horizontal scales equal to the horizontal scale of the map. This section should be fastened so that its horizontal lines are perpendicular to the extension of the strike line (BAC). The intersection of the strike line (·BAC) projected onto the 600-feet altitude line on the cross-section grid should be marked (point E). Through this point draw a line dipping 25° east on the cross section (line GEF). This represents the dip of the formation contact surface. A mark should be placed on the map at every point where the strike line of 600 feet altitude intersects the 600-feet contour line. The formation contact line will pass through each of these map points. Draw a strike line parallel to BACE and passing through the point where the dip line intersects the 500-feet altitude horizontal line in the section. Wherever this strike line of 500-feet altitude intersects the 500-feet map contour, place a mark to indicate the position of the formation contact line. Repeat this process for each contour. Complete the outcrop trace by connecting the marked outcrop points, being sure to use the rules of the V's where applicable. Points can be interpolated if necessary by using intermediate contour

FIG. 5–4 Method of extending formation contacts from outcrops where dip is known and constant.

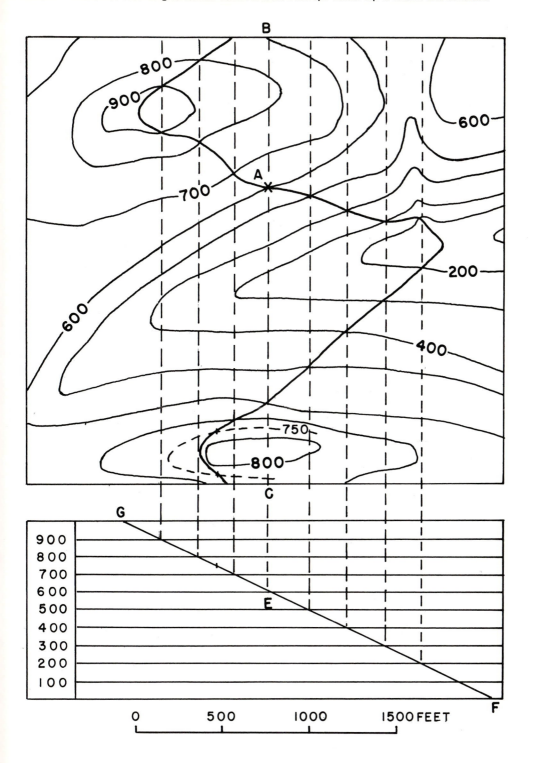

intervals. The outcrop of the contact can thus be precisely located between marked contours on the map (notice the interpolated 750-feet contour at the bottom of the map of Figure 5–4). If the subface outcrop of a formation is plotted and the formation thickness is known, then the formation thickness can be plotted on the structure section, and the outcrop of the superface projected from the structure section to the map.

In order to plot accurately the outcrop of a formation contact line in regions of very gentle dip, it is sometimes desirable to exaggerate the vertical scale of the cross section (see Chap. 2). The precise horizontal location of the outcrop can be more easily determined in this manner if the terrain is flat. As a word of caution, it should be noted again that formation thickness cannot be measured directly in such an exaggerated section, because the scale changes in different directions in the plane of the cross section.

Changing Dip / The methods described in the first part of this chapter apply only to strata with uniform dip—that is, to a homocline in which a bedding surface can be considered a plane.

The cross-section method of Figure 5–4 can be extended to situations where a curved bedding surface intersects the topography. The outcrop trace still follows the locus of points where the bedding and ground surfaces are at the same elevation (see Fig. 5–5).

Two closely spaced structure sections, such as those along lines AB and CD in Figure 5–5, will not coincide unless the folds have no plunge. If closely spaced structure sections differ only slightly, outcrop traces can be interpolated between the lines of section. For example, bedding surfaces AB and CD are coincident at point Y, shown in Figure 5–5, where sections AB and CD have the bedding trace projected from the respective lines of section onto a single vertical plane. The outcrop trace clearly is at elevation 500 feet on the map directly above point Y. At point X the outcrop trace is between 640 and 760 feet elevation, so on the map directly above point X in the section, point X' is plotted at 700 feet elevation proportional to the map location of the 700-foot contour as approximately half way between lines AB and CD.

A map of outcrop patterns can also be prepared if the altitude and configuration of the bedding surface are known well enough for a structural contour map to be prepared (Chap. 6). This situation could occur where dipping coal strata are poorly exposed, and an approximate structural contour map can be made on the basis of drilling results. The outcrop of coal can be determined for strip mining purposes by superposition of the structural contour map on the topographic map. Points at which topographic contour lines cross structural contour lines of the same altitude are located on the line of outcrop. A line connecting these known outcrop points completes the line of outcrop. The rules of the V's should be applied in drawing such outcrop lines.

FIG. 5–5 (opposite page)   Mapping formation contacts for beds with changing dip.

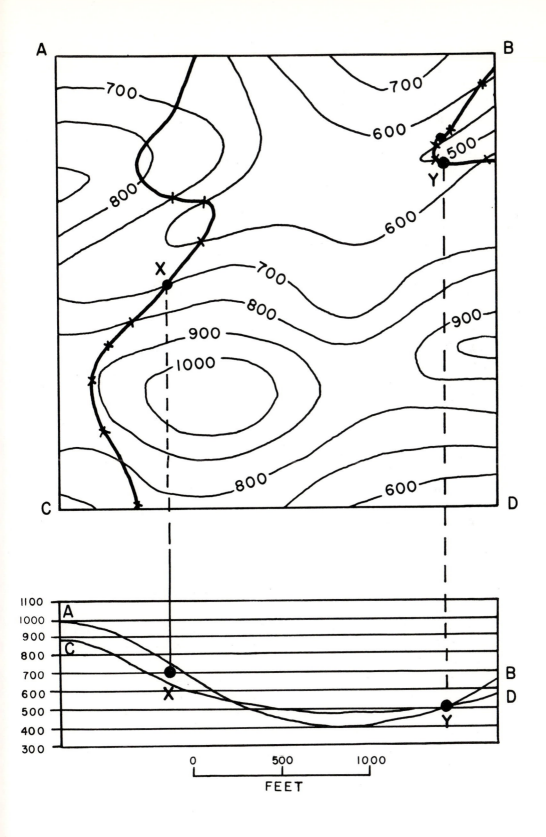

Exercises /

137. In a region with a uniform dip of 15°S80°E, the superface of a shale formation crops out at point A in Figure 5–6. A limestone formation occupies the area between points A and B. Calcitic shale occurs between B and C, and dolomite is between C and E. Sandstone crops out in the area east of E. Complete the formation contact lines, and determine the respective thicknesses of the limestone, calcitic shale, and dolomite.

138. The dip is uniformly 25°N35°W in Figure 5–6. Point A marks the subface of a sandstone and the superface of a limestone. Point B is the subface of the limestone and the superface of a shale. Point E is the subface of the shale and superface of a dolomite. Point C is the subface of the dolomite and superface of a siltstone. Draw the formation contact lines throughout the map. Determine the thickness of the limestone, shale, and dolomite.

139. In a region with a uniform dip of 20°S35°W, the superface of a dolomite crops out at point E in Figure 5–6. A sandstone formation crops out in the area between points B and E. A calcitic shale lies between B and C. Limestone crops out between A and C. Siltstone crops out in the area southwest of A. Complete the formation contact lines, and determine the respective thicknesses of the sandstone, calcitic shale, and limestone.

140. In Figure 5–6 the dip is homoclinal 13°N80°E. Point A marks the subface of a limestone and the superface of a shale formation. Point B is the superface of the limestone and the subface of a sandstone. Point C is the superface of the sandstone and the subface of a dolomite. Point E is the superface of the dolomite and the subface of a siltstone. Draw the formation contact lines throughout the map. Determine the thickness of the limestone, sandstone, and dolomite formations.

141. The coincident superface of a dolomite and subface of a limestone crops out at points C, E, and F of Figure 5–7. Point B marks the outcrop of the superface of the limestone and subface of a shale. Point A locates the outcrop of the superface of the shale and the subface of a sandstone. What is the dip of the beds? Complete the formation contact lines. What are the respective thicknesses of the limestone and shale?

142. Points A, C, and F of Figure 5–7 are on the superface of a limestone and the subface of a shale. Point E is on the subface of a limestone and the superface of a dolomite. Point B marks the contact between the shale and a sandstone. What is the dip? Complete the formation contact lines. What are the respective thicknesses of the shale and limestone?

143. The coincident superface of a shale and subface of a sandstone crops out at points B, E, and F of Figure 5–7. The subface of the shale and superface of a dolomite crops out at point C. Point A indicates the coincident outcrop of the superface of the sandstone and the subface of a limestone. What is the dip of the beds? Complete the formation contact lines as they would be shown on the map. What are the respective thicknesses of the sandstone and shale?

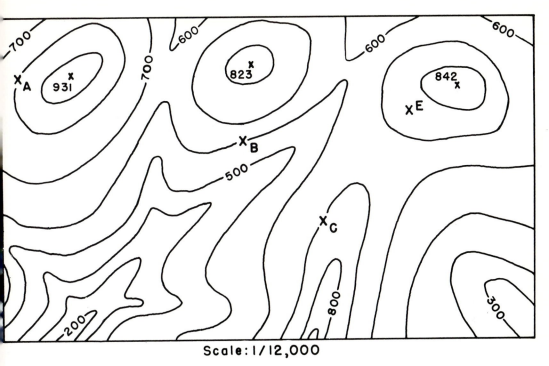

FIG. 5—6 Map for Exercises 137—140, 153, and 154.

FIG. 5—7 Map for Exercises 141—144.

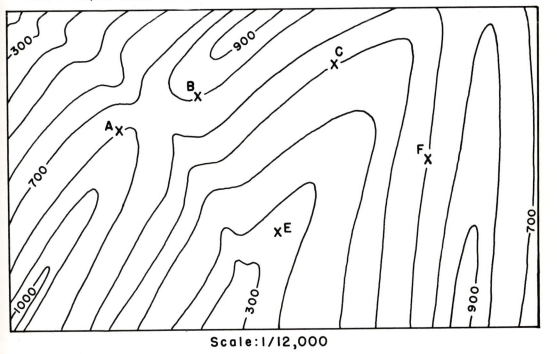

144. Points B, C, and F of Figure 5–7 lie along the outcrop of the subface of a gypsum bed in contact with dolomite. Point A marks the contact between the dolomite and a sandstone. Point E is at the contact of the sandstone and a limestone. What is the dip? Complete the formation contact lines. What are the respective thicknesses of the dolomite and sandstone?

145. The following stratigraphic succession of formations is recorded in the region of Figure 5–8:

Dolomite	800 feet
Sandy limestone	100 feet
Limestone	250 feet
Sandstone	130 feet
Shale	150 feet
Dolomite	900 feet

Point B is an exposure of the contact between the sandstone and shale. The regional dip is 10°N70°E. Plot the formation contacts that crop out in the area.

146. The sequence and thickness of formations are the same as those listed in Exercise 145. The contact between the sandstone and shale crops out at point A of Figure 5–8. The regional dip is 12°S18°W. Plot the formation contacts that are exposed in the area.

147. The following succession of formations is recorded in the region of Figure 5–8:

Shale, brownish gray	900 feet
Gypsum	125 feet
Limestone	225 feet
Sandstone	120 feet
Dolomite	160 feet
Shale, dark gray	900 feet

Point B is on an exposure of the contact between the sandstone and the dolomite. The regional dip is 8°N60°E. Plot the formation contacts that crop out in the area.

148. The sequence and thickness of formations are the same as listed in Exercise 147. The contact between the sandstone and dolomite is exposed at point A of Figure 5–8. The regional dip is 10°S23°W. Plot the formation contacts that are exposed in the area.

149. The basal contact of a sandstone is at an angular unconformity that is exposed at point A in Figure 5–9. The dip of the beds above the unconformity is 17°N38°W. The superface of the sandstone above the unconformity and the subface of a dolomite is exposed at point B. Point C is the outcrop of the dolomite superface and the subface of a calcareous shale. The top contact of a sandy shale and basal contact of a limestone, both older than the unconformity, is located at point E. The strata below the unconformity dip 33°S2°E. A narrow quartz vein, which developed before the formation of the unconformity and after deposition of the last beds below the unconformity, crops out at point F where it dips 57° due W. Assuming that the dips are constant, plot the outcrops of the unconformity, the formation contacts, and the vein. Determine the thicknesses of the sandstone and the dolomite.

150. The basal contact of a conglomerate is at an angular unconformity which is exposed at point A in Figure 5–9. The dip of the beds above the unconformity is 15°N35°W. The superface of the conglomerate above the unconformity and the subface of a sandstone is exposed at point B. Point C is an outcrop of the superface of the sandstone and the subface of a calcitic shale. The contact of a limestone and a dolomite below the unconformity is located at point E. The beds below the unconformity dip 30°S10°E. A narrow basalt dike, which developed before the formation of the unconformity and after the deposition of the last beds below the unconformity, crops out at point F where the dike dips 60°N85°W. Assuming that the dips are constant, plot the outcrops of the unconformity, the formation contacts, and the narrow dike. Determine the respective thicknesses of the conglomerate and sandstone.

151. A fault plane that dips 30°S63°E crops out at point B in Figure 5–9. At point C the contact between the subface of a shale and the superface of a dolomite dips 15°N5°W. The subface of a siltstone and the superface of a limestone crops out at point A with a dip of 25°S65°W. The subface of the limestone and superface of a sandstone crops out at point E. A narrow dike which formed after the fault has a dip of 55°N45°E at point F. Assuming that all the dips are constant, plot the outcrops of the fault, the formation contacts, and the dike. Determine the thickness of the limestone.

152. A fault plane which dips 32°S60°E crops out at point B in Figure 5–9. Point C marks the contact between the subface of a dolomite and the superface of a sandy shale which dips 18°S20°W. The subface of a sandstone and the superface of a limestone dip 26°S55°W and crop out at point A. The subface of the limestone and the superface of a calcareous shale crop out at point E. A narrow dike which formed after the fault has a dip of 62°N41°E at point F. Assuming that all of the dips are constant, plot the outcrops of the fault, the formation contacts, and the dike. Assume that the formations on one side of the fault are absent on the other side of the fault in the area of the map. Determine the thickness of the limestone.

153. Figure 5–10 is the base for a structural contour overlay to be used with Figure 5–6. It was drawn on the top of coal A from scattered exposures and drill core data. Carefully trace Figure 5–10 onto onionskin paper and place the onionskin as an overlay on Figure 5–6. Then plot on the overlay the location of a proposed strip mine that will strip all of the outcrop of coal A in the area. Coal B is another commercial coal that lies 100 feet stratigraphically above coal A. Plot the outcrop of coal B.

154. A geologist working with aerial photographs in the area covered by Figure 5–6 finds that measurements with a stereocomparograph show point A to be 750 feet altitude, point B to be 600 feet altitude, and point C to be 750 feet altitude. Assume that the altitudes of the three points on Figure 5–6 are correct as mapped and that the apparent difference in altitude results from tilt of the airplane from which the photograph was taken. What was the axis of tilt and the amount of tilt of the aircraft from which the photographs were taken?

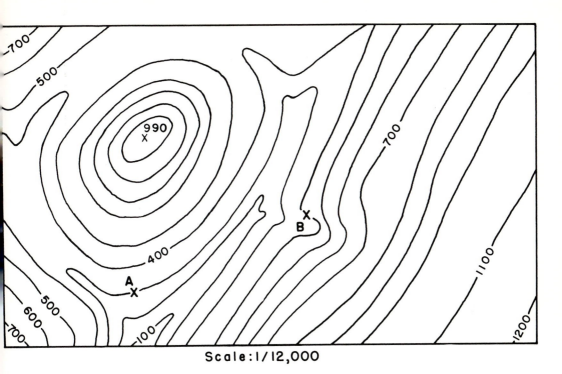

Scale:1/12,000

FIG. 5–8 Map for Exercises 145–148.

FIG. 5–9 Map for Exercises 149–152 and 160–162.

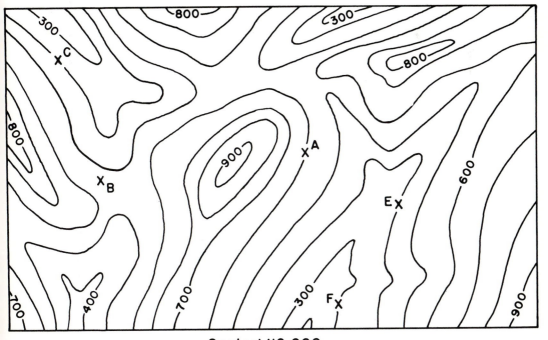

Scale:1/12,000

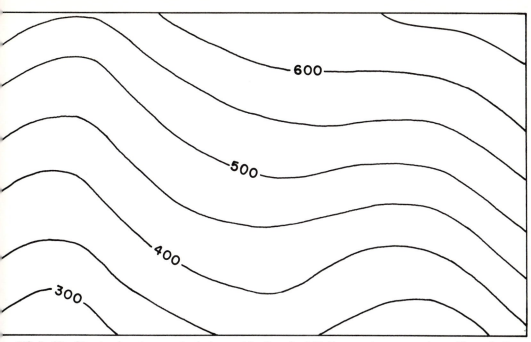

FIG. 5—10 Structural contour overlay to be used for Exercise 153. Trace and use as overlay on Figure 5–6.

FIG. 5—11 Structure section for use with Exercises 160 and 162.

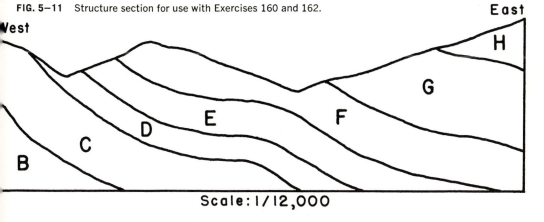

Scale:1/12,000

FIG. 5—12 Structure section for use with Exercises 161 and 162.

Scale:1/12,000

155. Solve Exercise 33 by the three-point method.
156. Solve Exercise 34 by the three-point method.
157. Solve Exercise 35 by the three-point method.
158. Solve Exercise 36 by the three-point method.
159. Solve Exercise 37 by the three-point method.
160. Figure 5–11 is a structural section across the bottom of Figure 5–9. Assume that the strike is north-south, and plot the formation outcrops.
161. Figure 5–12 is a structural section across the top of Figure 5–9. Assume that the strike is north-south, and plot the formation outcrops.
162. Figure 5–11 is a structural section across the bottom of Figure 5–9, and Figure 5–12 is a structural section across the top of Figure 5–9. Assume that the structural differences in cross sections result from uniform rates of change in a north-south direction. Plot the approximate positions of the formation contact lines, making a geologic map of Figure 5–9.

References

Donn, W. L., and Shimer, J. A., 1958, Graphic methods in structural geology: New York, Appleton-Century-Crofts, Inc., chapter 11 on dip and strike from three points, p. 74–83.
Hughes, R. J., Jr., 1960, A derivation of Earle's formula for the calculation of true dip: Jour. Geological Education, v. 2, no. 1, p. 43–48.
Screven, R. W., 1963, A simple rule of V's of outcrop patterns: Jour. Geological Education, v. 11, p. 98–100.

CHAPTER 6

Structure Contours and Isopachs

Geologists often seek to relate numerical values for data to the geographic location of the data points. One of the most useful ways to study areal trends of data is to construct isopleth maps (sometimes called isoline maps or isogram maps). *Isopleths* are loci of points that have equal numerical value and are interpolated from the plotted locations of the sample points and their respective numerical values. The most common geologic example is the construction of contour lines of equal elevation for *topographic maps*. Other types of isopleths include *structure contour maps* and *isopachous maps*. Structure contours represent connected points of equal altitude on a geologic surface such as a formational contact, a coal bed, or a fault surface. An isopachous map shows the thickness of a specified stratigraphic unit throughout a geographic area.

Procedure for Drawing Isopleth Maps

The familiar principles employed to plot contours on topographic maps are similar to those used to prepare all kinds of isopleth diagrams. The numerical value for each known point is first plotted at its proper location on the map or diagram. Sketching of the isopleths is then accomplished by interpolation between each pair of known points. Isopleths should be sketched lightly first, starting either with the highest or lowest line that will appear on the map, and then finished by rounding out any sharp curves not based on actual data.

There is no one best way for determining the position of specific isopleths on a map. A correct isopleth map is one that is consistent with assumptions used as guidelines for constructing it. Rettger (1929) named three distinct types of procedures for drawing isopleths: mechanical contouring, parallel contouring, and equi-dip contouring. Figures 6–1 through 6–5 illustrate different isopleth maps with the same data or control points.

Mechanical Contouring / In mechanical contouring it is assumed that the slope is uniform along lines connecting control points, so isopleths are positioned in arithmetic proportion. The data points of Figure 6–1 should be connected by straight lines to form a series of irregular triangles. The intersection of a specific isopleth with such a connecting line is determined by proportional interpolation between each pair of control points. For instance, consider preparing a map with an isopleth interval of 20 units using the two control points with values of 240 and 163 units at the right side of Figure 6–1. There must be one isopleth passing through the 240-units point and three other isopleths equally spaced between the two control points, except for the 180 isopleth, which will be proportionally closer to the 163-units point. The 180 contour will be 17/77 of the distance from the 163-units site toward the 240-units point. By using this method, isopleth points are obtained between each pair of control points, and those points with the same value are connected to form isopleth lines. Figure 6–2 is an isopleth map prepared from the data of Figure 6–1 using mechanical contouring technique.

Topographers generally use mechanical contouring to prepare a topographic map. The topographic surveyor deliberately places control points at topographic summits, along ridge crests, in stream valleys, at the branching of valleys, and especially at positions of significant break in slope. Consequently the slope (elevation change per unit horizontal distance) and contour spacing are essentially uniform between topographic control points. Mechanical contouring is ideally suited for this situation. Once the control for major topographic features is established, a careful topographer will alter the simple mechanical contouring by sketching minor topographic deviations from the simple mechanical pattern of perfectly regular slopes.

In most other isopleth situations the geologist cannot scan the area to pick out control points at the exact positions of "highs" and "lows" of the population values or the sites of abrupt changes of isopleth gradients. Instead he can use only those data that are obtained by whatever sampling process determines the location of the control points. Consequently there is considerable uncertainty about the true direction of trends or the positions of actual maxima or minima for isopleths in the region mapped, unless abundant sample points are very closely spaced.

Parallel Contouring / In parallel contouring the isopleths are drawn as nearly parallel with one another as possible. The spacing between isopleths can be made variable in order to attain the goal of parallelism. Figure 6–3 is an isopleth map developed by parallel contouring of the data from Figure 6–1.

Parallel contouring is less conservative than true mechanical (arithmetic) spacing

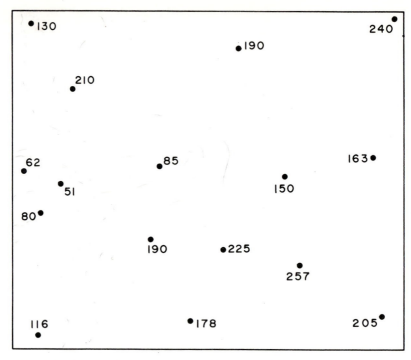

FIG. 6–1 Numerical values for data points that occur in all isopleth maps of Figures 6–2 through 6–5. Horizontal scale not specified.

FIG. 6–2 Isopleth map drawn by using mechanically spaced contouring of data points contained in Figure 6–1.

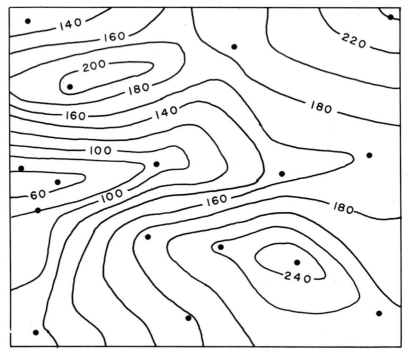

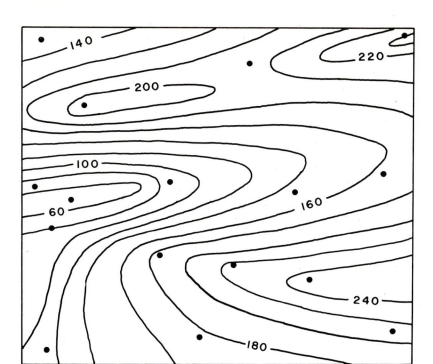

FIG. 6—3   Isopleth map drawn by using parallel contouring of data points contained in Figure 6—1.

FIG. 6—4   Isopleth map drawn by using equi-dip contouring of data points contained in Figure 6—1.

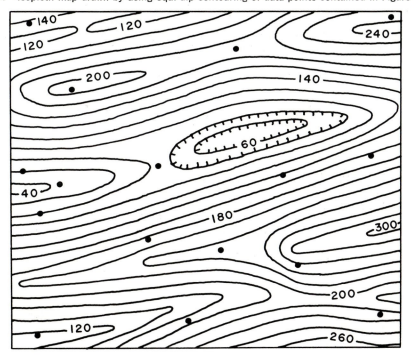

of contours. Once a regional trend is established, minor structures are interpreted to blend into the regional linear trend. Parallel contouring has a tendency to indicate structures where in mechanical contouring merely a flattening of dip might be shown; such structures tend to die out if extended into the unknown.

Equi-dip Contouring / In equi-dip contouring the isopleths are spaced equal distances apart, the distance being determined by the dip between two relatively close control points or by a previously determined regional dip. The dip or spacing is ordinarily the steepest or smallest found in the area being mapped. Figure 6–4 is an equi-dip isopleth map prepared from the data points in Figure 6–1.

Equi-dip contouring is the most radical of the three methods. It leads to considerable inference regarding possible or probable structures. Instead of contouring flat dip areas as such, reversals of dip and new structures are mapped. Compared to the mechanically contoured map of Figure 6–2, the equi-dip isopleth map (Fig. 6–4) has several additional structures: isopleth highs at two places on the right margin, lows near the upper left and lower left corners, and an isopleth depression near the center of the map. Equi-dip contouring is a useful indicator of the maximum number and amplitude of probable isopleth "high" and "low" areas, because it assumes that all slopes are as steep as the maximum slope between any control points.

Random Points from an Irregular Surface / Figure 6–5 is an isopleth map showing a hypothetical "true" configuration of an irregular surface. Suppose this is the true elevation of a bedrock surface buried unconformably beneath Pleistocene till. Randomly located test borings at the marked sites would yield topographic elevations as indicated in Figure 6–1. Even though the true configuration of the irregular surface is as shown in Figure 6–5, it is reasonable to interpret its configuration by the isopleth maps of Figures 6–2, 6–3, or 6–4, depending on the assumptions made for the "proper" contouring procedure.

The difficulty with such control points is that they are not placed at significant localities with respect to isopleth summits, depressions, ridges, troughs, or significant breaks in slope. Yet many geologic isopleth maps, particularly those based on subsurface control, are based on some sampling locality scheme not related to the true configuration of the surface: random sampling, sampling on a rectangular grid, or simply using all data available even if it was not located by predetermined pattern. Such maps have inherent lack of reliability compared to the more familiar topographic map based on control points carefully chosen from scanning the true configuration of the irregular surface.

As spacing of control points becomes closer, uncertainties diminish so that mechanical contouring becomes a progressively better approximation to the calculus of the true configuration of the surface.

Tendency to Interpret Regional Trends / Geologists naturally want to discover trends in the isopleths of a region, and one is tempted to interpret their existence even when none is present. Dodd, Cain, and Bugh (1965) described an interesting experiment in which north-south and east-west map coordinates were selected at random and then random numbers were assigned to these geographic "control" points. Use of random number tables ensured that there were no significant deviations from random variation. These points were contoured assuming they represented four geologic situations: simple mechanical contouring, contouring of terrain into a reasonable topographic pattern, as structural contours, and as a deviation map from a mathematical equation trend surface. In all cases it was possible to contour the data points to yield geologically reasonable maps, even though the maps were based on random data that had no true trend of variation. Beware of reading too much into your maps!

Structure Contour Maps

Structure contours connecting points of equal altitude on a geologic surface are drawn as separate structure contour maps or are overprinted on geologic maps to show the areal variation of the vertical distance of a geologic horizon above a datum surface (elevation of the geologic horizon, if the datum is sea level). Structure contours are sketched from points of known elevation on the key horizon or bed being contoured.

FIG. 6–5 Isopleth map of surface from which the numerical values in Figure 6–1 were obtained by random sampling for geographic location of data points. Using just the numerical values, and interpolating trends by various procedures resulted in the maps of Figures 6–2 through 6–4. The interpreted map may show little resemblance to the true configuration of the irregular surface.

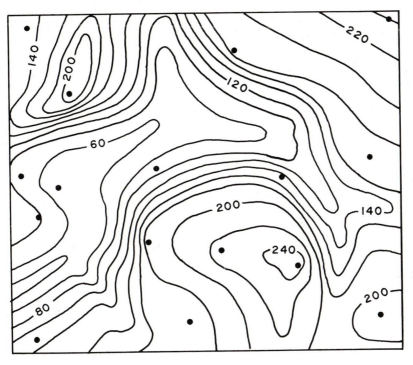

These points of altitude are determined from drill holes, mine openings, or the projection of surface outcrops. Points determined in drill holes and mine shafts can be used directly, along with the elevation of the outcrop trace of the key horizon. Other elevations for structure contouring can be computed from geologic map data if the thickness and dip of the beds in the area are known. The structure contour interval should normally be two or three times the accuracy limits of the data contoured. Of course, structure contours cannot show structures smaller in amplitude than the contour interval. Although structural contouring is largely a mechanical process, if data points are closely spaced, still wise geological judgment should be used at all times. Strata deform into rounded flexures, so in general the configuration of structure contours consists of smoother curves than the irregularities of topographic outcrops, unless the structure contour pattern is cut by a fault.

It is conventional to use solid structure contour lines to represent surfaces below ground, whereas structural contours consisting of long dashes are used for a geologic surface that has been eroded and that should project in space above the present ground surface. Structural depressions, such as a structural basin or a syncline, are shown by hachured structure contours, analogous to hachured contours for topographic depressions. Structure contours may have negative values if the contoured surface is below sea level or other datum. All structure contours are closed curves, although closure may be completed outside the area being mapped. In this respect they are comparable to topographic contours.

Vertical faults, such as the one shown in Figure 6–9, show an offset in structural contours at the fault line. In a normal fault the horizon contoured is not present continuously across the map view of the fault, and the structure contours also show a gap (Fig. 6–6, map *a*). In a map view of a reverse fault, the structural contours on opposite blocks of the fault overlap as in Figure 6–6, map *b*. The portion of the relatively underthrust block on the underside of the overlapping area has its structural contours

FIG. 6–6 Methods of showing nonvertical faults on structure contour maps.

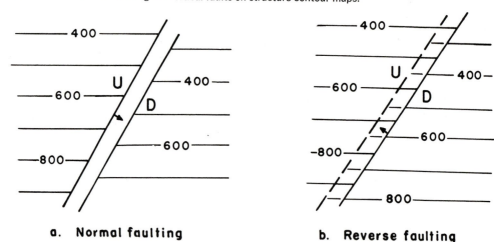

a. **Normal faulting**   b. **Reverse faulting**

represented by short dashed lines. The throw (vertical component of dip separation) is 150 feet in both diagrams in Figure 6–6. The heave (horizontal component of dip separation) is the width of the gap or band of overlap, measured in the direction of the arrow in Figure 6–6. It is customary to mark the upthrown block of a fault with the letter "U" and the downthrown block with "D." Faults are interpreted on a structure contour map if the elevation of the contoured horizon varies too abruptly to be explained by reasonable folding. Faults are more obvious if they cut across the regional strike of the strata than if they are strike faults.

Structure contour maps made from well data are based on control points located independently of any certain knowledge about the positions of crests and troughs of folds. If well control is scattered and a pronounced regional trend of fold axes is known to be present, the steep-limbed folds can be sketched by the parallel contouring method. The map of Figure 6–3 is probably a good representation of near-concentric folds produced by a simple north-south compressive force. Comparable fold patterns occur in the Appalachian Mountains of West Virginia, Maryland, and Pennsylvania.

The equi-dip structure contour pattern of Figure 6–4 is close to that resulting from plunging chevron folds. These probably would not occur on large scale, but a map of steep-limbed similar folds would appear much like this.

If control points are very closely spaced relative to the wave length of the folds, then mechanical contouring with smooth deflections of isopleths is probably the best procedure. If no prominent set of sub-parallel fold axes occurs in the region, or if the beds dip very gently, mechanical contouring is better than parallel or equi-dip contouring.

Bishop (1960, p. 47–48) applied the name *interpretative contouring* to the method she considers best for regions with sparse control. These structure contours deviate somewhat from the mechanical form to indicate patterns that are consistent with a given set of values as well as the known (or supposed) structural habit of the region. The more sparse the data points, the more varied the possible interpretations.

If the geologist can do field work on outcropping structures, he can obtain structural contour control approaching the quality of that used for topographic isopleths. Fold crests and troughs can be located; with a knowledge of stratigraphic thickness of formations it should be easy to calculate the approximate elevation of the reference horizon, either buried in the ground or projected into space to its position before erosion. The direction of regional fold trends can be established by mapping. Elevation of the outcrop trace of the reference horizon provides additional data for structural contouring. Well and mine data add to this information. The resulting map should be a fairly accurate representation of near-surface structures.

Stratigraphic Thickness Maps

Isopleths can be drawn to show the thickness of a bed, formation, sill, or other tabular body throughout a geographic area. These are commonly referred to as *isopach maps,* and the individual lines are called *isopachs.* Control points are located independently from the true position of maxima and minima, so these maps are interpretive estimates of the true thickness configuration. Unusually thinned portions of the isopached unit are indicated by hachured isopachs, and portions thicker than the rest of the unit are shown by closed isopachs comparable to the closed contours indicating a hill on a topographic map. The isopach interval should be two or three times the accuracy limits of the data used to prepare the map. In regions where the control points are widely spaced and thickness data suggest considerable variation from simple tapering of a planar wedge, a rather large isopach interval should be used to avoid the implication that thicknesses have been interpolated quite accurately over the map area. Isopach maps of units whose thickness results chiefly from continuous depositional accumulation usually have gradual and simple patterns of isopach curvature. Isopach maps of units bounded above or below by unconformities may exhibit intricate isopach patterns resulting from topography of the ancient erosion surface. Of course, only points representing complete thickness of the formation should be plotted for isopach control. Geologic literature contains numerous measured sections of portions of formations; only sources that clearly label the top and bottom of the formation should be used for thickness data.

Several types of information yield approximate thickness of a formation, so different types of thickness maps are possible. Specific kinds of thickness maps are called isopachous, isochore, and convergence maps.

Isopachous Maps / An *isopachous* map is one that shows the true thickness of the formation or other geologic unit, measured normal to the bounding surface. Valid data include measured sections of outcrops whose thickness has been calculated by the methods of Chapter 3, well thicknesses obtained in a vertical hole drilled through essentially horizontal strata, or well data that have been mathematically corrected to yield true thickness. The term *isopachous* map is used here for one showing true thickness normal to bedding or other bounding surface, as distinguished from *isopach* map used for thickness maps based on a variety of types of data.

Isochore Maps / Thickness information can also be obtained from drilling data. If the angle between the drill core axis and the bedding surface is known, then the true thickness can be calculated by the formula:

Thickness = (Drilled distance) (Cos angle between core axis and bedding). (54)

In cable tool or rotary drilling procedures in which the top and bottom of the formation are identified by examination of the pulverized rock, the drilled thickness of the formation is the drill hole footage from the top to the bottom of the formation. A map prepared from such apparent thickness data is called an *isochore map.* A variety of factors may produce discrepancies between isochore and isopachous data. In lines CD and EF of Figure 6–7 the drill hole itself is not vertical. A borehole survey can be run, at considerable expense, to measure the true attitude of the hole throughout its length, and vertical depth can be calculated as follows:

Vertical depth = (Drill hole distance)(Cos angle between hole and vertical). (55)

These calculations can be made in short increments along the length of the hole, yielding true vertical depths to formation tops and resulting in more accurate structure contour maps and better formation thickness data. In fact, some areas that have been interpreted as a series of gentle folds really have essentially flat strata, with the apparent folds resulting from failure of the geologist to make vertical depth corrections for lateral wanderings of the drill holes.

Isochore thickness may exceed true thickness if the vertical hole penetrates dipping beds (line GH in Fig. 6–7). This can be corrected by the formula:

True thickness = (Drilled thickness in vertical hole)(Cos dip angle of beds). (56)

The effect of this correction is slight for gentle dips. Apparent thickness exceeds true thickness by 0.4% if the dip is 5°, by 1.5% if the dip is 10°, by 6.4% if the dip is 20°, and by 15.5% if the dip is 30° (Weller, 1960, p. 646).

Of course, present thickness values do not represent thickness of the unit at the time of deposition, if there has been appreciable compaction. This generally is of little concern for structural geology purposes, although it is interesting stratigraphically. Of greater importance structurally is the problem of tectonic thinning (IJ in Fig. 6–7) or tectonic thickening (KL and PQ). Drag-folded units can cause unusual thickening in a well penetrating soft shale (PQ), but the presence of such a complication is very difficult to detect in well data except by actual examination of drill cores, which are costly to obtain. Whenever possible, a geologist should avoid using tectonically distorted thickness values to prepare isochore and isopachous maps.

FIG. 6–7 Factors that may cause drilled thickness to deviate from true thickness.

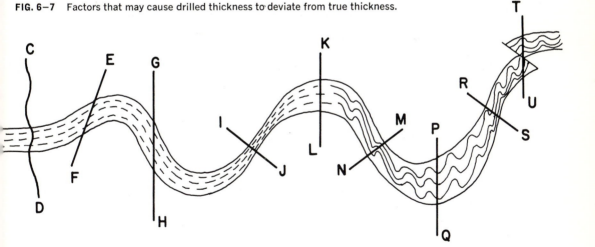

Convergence Maps / If the well elevation of the superface and subface are recorded, then the difference of these two elevations is a measure of the thickness of the formation. A formation thickness map constructed this way is a *convergence map* (Levorsen, 1927, p. 680; Krumbein and Sloss, 1963, p. 439). If the formation has appreciable dip, the convergence thickness values are somewhat greater than the true thickness. The advantage of convergence maps is that well data are generally summarized as depth to formation contacts, and these figures can be easily converted to convergence values to produce a thickness map.

Uses of Thickness Maps / Some geologists make no distinction among isopachous, isochore and convergence maps. Others identify isopachous maps separately as described above, but use isochore and convergence maps as interchangeable names and concepts. Usually all three types of maps are similar enough that they can be used for the same purposes, but it is occasionally necessary to distinguish the three slightly different map concepts.

Isopachous and isochore maps are often used as overlays on structure contour maps of the superface of a bed in order to compute points of altitude on the subface of the bed. These points are then used to construct a structure contour map of the subface. The correctness of such a subface structure contour map will depend on the geographic spacing of the known points of thickness and the validity of the assumed constant rate of thickening between control points. For instance, isopachs are often shown as straight lines when they are based on information from a limited number of drill holes, although it is known that this is seldom the true picture. Actually the isopachs are more often curved in irregular patterns. Nevertheless, even straight isopachs are very useful in predicting directions of migration of structures with depth.

Exercises / Exercises 163–166 are based on areal geology maps showing the contacts between a series of sedimentary formations and the altitudes of a number of points along the contacts. In all four exercises formation A is the oldest formation studied. All of the formations except A are of constant thickness in the mapped areas. The thickness of formation A varies a great deal as shown by the thickness recorded at a control point near each corner of the maps. It is desired to:
1. Prepare a structure contour map of the superface of formation A.
2. Prepare an isopachous map of formation A.
3. Prepare a structure contour map of the subface of formation A.
4. Describe the structures on the superface and subface of formation A, and compare the structures on the two subfaces.

STEP 1. First convert the altitudes of the points on the surface formation contacts into altitudes of points directly beneath on the superface of formation A. In order to compute the altitude of the superface of formation A, subtract the vertical distance between the known elevation point on the formation contact line and the superface of formation A from that known elevation value on the outcrop map. For this exercise assume that the dips are gentle enough and the wells are nearly enough vertical so that stratigraphic thickness may be used as the vertical distance between the superface of formation A and the overlying outcrop of formation contacts. Repeat this procedure for each point of known altitude on the formation contact lines. The result is an array of points of known altitude on the superface of formation A. Mechanically contour these points to show the configuration of the superface of formation A, using a structural contour interval of 100 feet.

STEP 2. For just four control points the best fit of isopachs will be a series of straight lines whose position is determined by the stratigraphic thickness of formation A at the control points. The general method of locating points on the isopachs is similar to that of locating points through which topographic contours pass. Series of points through which isopachs will pass can be located along lines connecting the four wells along the map borders. The distance along such a line from one well to a specific isopach can be determined from the following proportion:

$$\frac{\text{Distance between specific isopach and first well}}{\text{Differences in stratigraphic thicknesses at specific isopach and first well}} = \frac{\text{Distance between first and second wells}}{\text{Differences in stratigraphic thicknesses at first and second wells}}. \quad (57)$$

After the locations of the specific points through which isopachs pass are determined, connect those points with the same thickness by means of straight line isopachs. Draw the isopachous map on a thin sheet of paper that can be used later as an overlay on the structure contour map of the superface of formation A. Use an isopach interval of 20 feet.

STEP 3. The structure contour map of the subface of formation A should be constructed on the same sheet as the isopachous map. Use different colored pencils for the structure contours and isopachs. First fasten the isopachous map as an overlay on the structure contour map of the superface. Altitude points on the subface are located by marking every crossing of a superface contour by an isopach. The points thus established are given altitude values equal to the superface contour altitude minus the isopach value at that point. It is assumed that the dips are gentle enough that isopach values can be used as isochore values. The computed altitude points should be left on the structure contour map. These altitude points are then mechanically contoured, using a structural contour interval of 100 feet.

STEP 4. The structures on the superface and subface of formation A should be located and described by listing their trends, symmetry, dips of their flanks, the directions and amounts of plunge, and the amounts of closure. The structures on the superface and subface should be compared by noting the appearance and disappearance of structures and the migration of structures with depth. Changes in the amplitudes of the structures should also be recorded.

163. The geologic map of Figure 6–8 is the reference for this exercise. The following stratigraphic sequence is present in the area:

FORMATION	THICKNESS
G	not determined
F	55 feet
E	175 feet
D	90 feet
C	270 feet
B	210 feet
A	Well 1—520 feet
	Well 2—370 feet
	Well 3—150 feet
	Well 4— 80 feet

164. The geologic map of Figure 6–8 is the reference for this exercise. The following stratigraphic sequence is present in the area:

FORMATION	THICKNESS
G	not determined
F	65 feet
E	185 feet
D	90 feet
C	270 feet
B	240 feet
A	Well 1—500 feet
	Well 2—350 feet
	Well 3—150 feet
	Well 4—100 feet

165. The geologic map of Figure 6–9 is the reference for this exercise. The stratigraphic sequence of the area is as follows:

FORMATION	THICKNESS
H	not determined
G	140 feet
F	80 feet
E	160 feet
D	110 feet
C	170 feet
B	90 feet
A	At northwest corner of map— 40 feet
	At northeast corner of map— 95 feet
	At southwest corner of map—193 feet
	At southeast corner of map—257 feet

Assume that the fault is vertical.

166. Use the geologic map of Figure 6–9 for this exercise. The following stratigraphic sequence occurs in the area.

FORMATION	THICKNESS
H	not determined
G	170 feet
F	80 feet
E	60 feet
D	130 feet
C	190 feet
B	90 feet
A	At northwest corner of map— 50 feet
	At northeast corner of map— 95 feet
	At southwest corner of map—176 feet
	At southeast corner of map— 238 feet

The fault is vertical.

Additional exercises can be developed to utilize the maps of Figures 6–8 and 6–9. Exercise 164 was modified from Exercise 163 by changing the thickness of formations B through F. Thickness of formation A is varied by changing the thicknesses encountered in the respective wells. Of course, it would be possible to develop exercises with two formations varying in thickness over the area.

FIG. 6–8 Areal geologic map for Exercises 163 and 164. Scale 1 inch equals 1 mile.

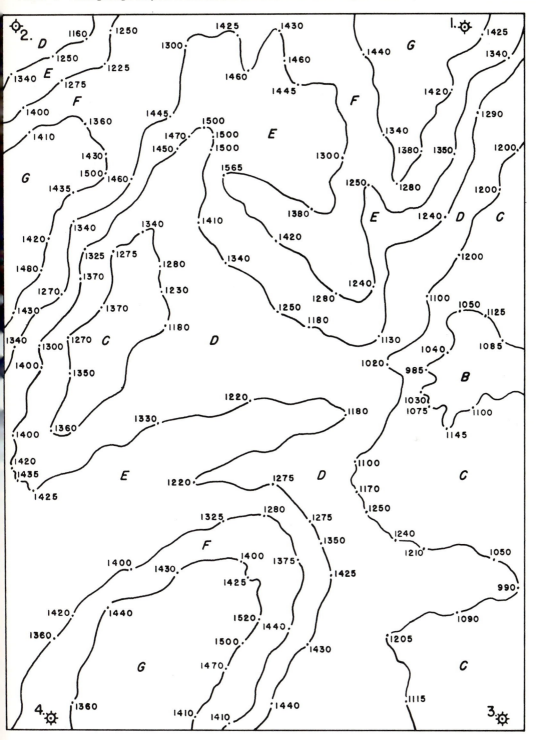

FIG. 6–9 Areal geologic map for Exercises 165 and 166. Scale: 1:24,000.

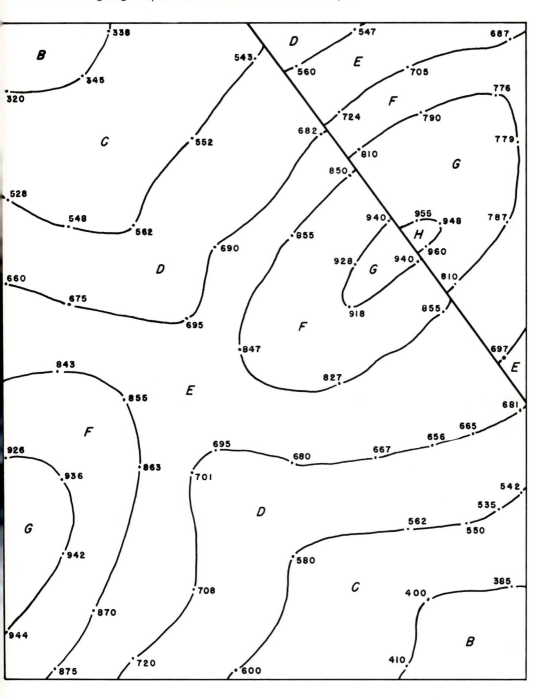

References

Badgley, P. C., 1959, Structural methods for the exploration geologist: New York, Harper and Bros., p. 70–151. (Structural contour and isochore maps.)

Bishop, M. S., 1960, Subsurface mapping: New York, John Wiley & Sons, 198 p. (See especially p. 33–98).

Dickinson, G., 1954, Subsurface interpretation of intersecting faults and their effects on stratigraphic horizons: Am. Assoc. Petroleum Geologists Bull., v. 38, p. 854–877.

Dodd, J. R., Cain, J. A., and Bugh, J. E., 1965, Apparently significant contour patterns demonstrated with random data: Jour. Geological Education, v. 13, p. 109-112.

Harrington, J. W., 1951, The elementary theory of subsurface structural contouring: Am. Geophy. Union Trans., v. 32, p. 77–80.

Krumbein, W. C., and Sloss, L. L., 1963, Stratigraphy and sedimentation: San Francisco, W. H. Freeman and Co., 2nd ed., 660 p.

LeRoy, L. W., and Low, J. W., 1954, Graphic Problems in petroleum geology: New York, Harper and Bros., p. 6–7, 135–139, 170–182. (Example of preparing isochore map and preparing structural contour maps from dips and strikes and from topography and areal geology.)

Levorsen, A. I., 1927, Convergence studies in the mid-continent region: Am. Assoc. Petroleum Geologists Bull., v. 11, p. 657–682.

Moore, C. A., 1963, Handbook of subsurface geology: New York, Harper and Bros., 235 p. (See especially p. 1–17, 37–54.)

Nevin, C. M., 1949, Principles of structural geology: New York, John Wiley & Sons, p. 68–71.

Reiter, W. A., 1947, Contouring fault planes: World Oil, v. 126, p. 34–35.

Rettger, R. E., 1929, On specifying the type of subsurface structural contouring: Am. Assoc. Petroleum Geologists Bull., v. 13, p. 1559–1561.

Weller, J. M., 1960, Stratigraphic principles and practice: New York, Harper and Bros., 725 p.

CHAPTER 7

Interpretation of Geologic Maps

A STRUCTURAL GEOLOGIST is often required to interpret the work of others, especially in large companies where more than one party may be in the field at the same time. More often, it is necessary to interpret published maps and even one's own work after the field season is over. So many field observations are recorded on geologic maps that the correct evaluation of structures shown on them is an important part of geologic success.

Maps as Interpretations

In all honesty, every geologic map is something of an interpretation in itself, because the bedrock of an area is seldom completely exposed, or if it is, the mapping geologist probably did not personally examine every square foot of the mapped terrain. The published geologic map portrays the areal distribution of rock types as envisioned by the mapper after he has spent considerable time examining the rocks in the region. It probably will always be possible to return to an area for more detailed mapping, perhaps on a larger scale base map, and to make improvements on previous maps. In this sense all geologic maps are "progress reports" of the opinion held by the mapping geologist at a particular time.

In order to separate fact from interpretation, detailed mapping operations often produce both an outcrop map and a geologic map. An *outcrop map* shows the distribution and shape of outcrops of lithologic units, along with measured data for specific places such as strike and dip of beds, plunge of lineation, and specimen or fossil collections. A *geologic map* groups these outcrops into bands or areas of formations and shows geologic structures such as fold axes, positions of known or suspected faults, along with summary markings of strike and dip, lineation, etc. Probable rock types in areas between actual exposures must be interpreted, and there is sometimes considerable personal judgment about what should be the formation assignment of a particular outcrop. Even the accompanying columnar section of rock types and their age assignment frequently reflects personal opinion of the geologist.

In summary, a geologic map is only as good as the combined field work and interpretative understanding of the geologist who made the map. The outcrop trace of formation contacts usually reflects to a great degree the structural interpretation favored by the mapper.

Age Relations

The relative and absolute ages of rocks help one interpret the information contained on a geologic map. The following relationships are useful in this regard:

1. Principle of superposition. Younger sediments overlie older sediments, unless the strata have been overturned by deformation.
2. Principle of physical continuity. Beds are generally continuously present throughout an entire area unless they have been eroded away. If certain beds are locally absent or terminated abruptly, suspect faulting, truncation by an unconformity, or lenticular and local deposition.
3. Rule of deformation. Deformed rocks were created before the time of deformation.
4. Rule of metamorphism. Metamorphic rocks were some type of sedimentary or igneous rock before metamorphism occurred. The type of rock before metamorphism frequently can be interpreted, along with the types and stages of metamorphism.
5. Rule of cross-cutting relationships. Rocks cut by an intrusion, fault, joint, or a replacement structure are older than the cross-cutting feature.
6. Age determination by fossils. Paleontologic evidence usually enables one to establish the age of strata and may allow one to demonstrate the existence of an unconformity between two stratigraphic units. Caution should be exercised about facies fossils and paleoecologic relations. Position of a fossil in an evolutionary sequence aids precision of age determination of the enclosing strata.
7. Absolute dating by radioactivity. This method allows one to relate rocks directly to the geologic time scale in years. Intrusions and metamorphic relations can be dated this way, as well as certain sediments.
8. Age determination by facies relationships. Intertonguing lithologic units are contemporaneous.

Multiple Hypotheses

It would not be much of an overstatement to say that all geologic knowledge is useful in geologic map interpretation. For this reason, little specific advice can be given except to make every interpretation through a conscious review of the character and *composition* of the rock section, the geologic *structures* that may be present, the *processes* to which they have been subjected, and the *stage* of development reached in response to the processes.

The scientific method of using multiple hypotheses (Chamberlin, 1897) should always be applied when interpreting geologic maps. The best hypothesis is generally the one that accounts for the most geologic features on the map. Often, but not always, the best hypothesis turns out to be the simplest.

Be alert to distinguish features that have a purely stratigraphic or depositional history from those of later structural origin. An example is that the width of an outcrop of beds may vary because of facies changes or changes in the thickness of a formation rather than because of faulting or changing dip. Topography also influences the width of outcrops.

Down-structure Method of Viewing Geologic Maps

Mackin (1950) has described what he calls the "down-structure method" for examining geologic structures in map view. His paper explains reliable methods for quickly interpreting relative displacements of faults or complex folding on geologic maps. The basic idea will be presented here, and the original article is recommended for more specific examples.

Most conventional geologic maps are framed by meridians and parallels, with printing and symbols arranged so that "north is up." This forces a standard orientation of the map relative to the ground and to regional geographic and geologic patterns. Most lines on geologic maps are traces of inclined planes or surfaces, the orientation of which is unrelated to the geographic framework of meridians and parallels. In areas of steeply dipping beds, topographic deflections of formation contact lines are small, and the major source of map patterns is folds or faults in the strata themselves.

If Figure 7–1 is oriented so that the observer views it in the direction of dip or plunge, the line S-S′ becomes a horizontal surface, the portion of the map toward the observer is the structure below the surface along the line of section, and the portion of the map beyond the line is the structure removed by erosion above the surface. In practice line S-S′ can be a straight edge or card that can be shifted at will, so long as it is kept normal to the direction of dip or plunge of those structures on which attention is focused. One can also dispense with the straight edge to see the structure in its entirety.

Figure 7–1 has been drawn with vertical axial planes, so that down-structure viewing produces nearly a true cross section of the folds. View the geologic map from the east, looking in the direction toward which the folds plunge. It will be seen in down-plunge view that the anticlines and synclines are directly visible without one's

FIG. 7–1 Diagrammatic map to illustrate several simple applications of the down-structure method. (From "The Down-structure Method of Viewing Geologic Maps" by J. Hoover Mackin, *Journal of Geology*, 1950, The University of Chicago Press.)

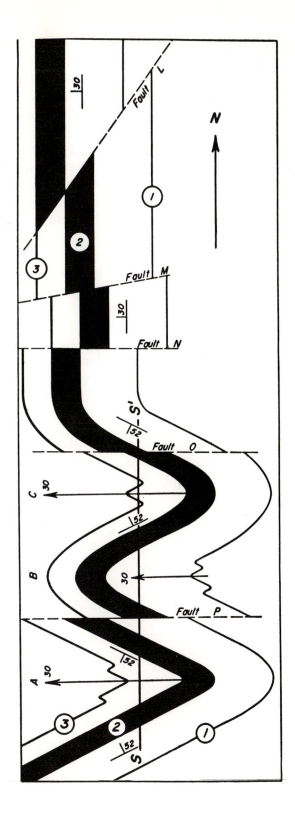

having to figure the individual age relations of the beds on the map. The view is a cross section in a plane normal to the axial plunge, rather than an exact vertical plane. The down-structure view shows that the folding is disharmonic, and it follows that the down-plunge view provides a closer approximation of the types of subsurface structures than does any single cross section based on downward extension of bedding attitudes measured along the line of section. The stratigraphic position of horizons characterized by subsidiary folding is evident, and these minor folds do not persist into the other less pliable strata. It is assumed that the kinds of structural features in individual beds and corresponding parts of folds are continuous down the plunge.

In the faulted terrain mapped in the northern part of Figure 7–1 one can readily see the relative displacement of the blocks by looking into the map down the *bedding plane.* Even if nothing is known or assumed concerning the direction or amount of movement on the faults, the down-structure view provides at a glance a correct picture of the stratigraphic throw of any horizon within the moved blocks. For example, in block N the stratigraphic throw equals the thickness of bed 2. The down-dip view of faulted beds may cause the faults to appear to have different dips than in a vertical cross section. Faults L and M are actually vertical, yet viewed down the dip of the bed they intersect the strata obliquely. In the inclined plane the fault traces really do not cross the beds at right angles, but instead have an oblique transect.

Figure 7–2 is a map of a plunging and faulted fold. The oldest beds are toward the center, so it is anticlinal. Since the oldest beds are generally on the upthrown block of

FIG. 7–2 Geologic map for applying down-structure method to plunging folds. Successively numbered formations are progressively younger.

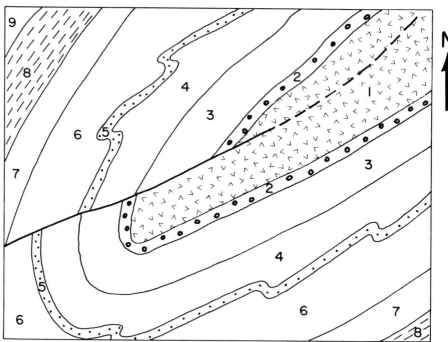

a fault, the fault appears to be vertical or steeply dipping with relative uplift of the south block. Now turn the map so that it is viewed toward the direction of plunge. The anticlinal nature of the fold is very apparent. The fault now can be interpreted as a high-angle reverse fault cutting one limb of the anticline. Bed 5 is an intricately folded unit that is incompetent compared with beds 2, 3, and 8.

Exercises /

[Exercises 167–187 are based on the hypothetical geologic map of Figure 7–3.]
The student should be able to answer most of the questions with the aid of his general geologic background. In Figure 7–3 it is assumed (a) that there has been only dip-slip movement (no strike-slip movement) and (b) that the high-angle faults are all the same age. The thin, solid lines are contour lines with elevations in feet as numbered. Thick, solid lines are faults. Formation contact lines are represented by dotted lines and pattern changes. A stream system is shown by solid lines of moderate width.

167. What is the major structure shown on the map?
168. One inch on the map is equivalent to how many feet in the field?
169. The fault blocks in the southwest rectangle are of what type or types?
170. What term is applied to the northeast-striking body of sediments through which passes the line separating the SE rectangle and the E rectangle?
171. Place arrows at the bottom of the map indicating true north and a magnetic declination of 6°W. Use conventional symbols. The map border is parallel to true north.
172. Which is older, the monzonite or gabbro? Give the reason for your answer.
173. What term is applied to the structure between the parallel faults in the NW rectangle?
174. In what area or areas would the presence of contact metasomatic deposits of copper be most likely? Explain.
175. What is the probable origin of the alluvial deposits near the center of the map?
176. What name may be applied to the granodiorite pluton at the wrap-around of the fold?
177. How many major unconformities are there, and where are they located? What is the evidence for your statement?
178. Account for the difference in the width of outcrop of the Silurian formation from place to place.
179. What is the term applied to the small body of sediments striking N10°W found in the E rectangle?
180. How is it possible to explain the difference in direction of the relative movement of each of the formation contact lines displaced by faulting in the S rectangle? (Remember it is assumed that there has been no strike-slip motion along the faults.)

181. Which is younger, the granodiorite or gabbro? Explain.
182. In what direction does the major structure plunge?
183. Assuming the granodiorite to be middle Tertiary in age, what is the geologic age of the low-angle fault in the NW rectangle?
184. Assuming no strike-slip, what is the amount in feet of the heave and the throw along the south fault in the SW rectangle if the dip of the Cambrian formation is 30°?
185. What is the name applied to the three plutons nearest the north margin of the map?
186. In what direction does the low-angle fault dip under most of the klippe in the NW rectangle?
187. Write a summary of the geologic history of the map area.

The remaining exercises in this chapter are more complex than the preceding. These next exercises are based on folios of the *Geologic Atlas of the United States* and *Geologic Quadrangle Maps*, both series issued by the United States Geological Survey. The aim is to familiarize the student with the types of structures that occur in several different geologic environments.

[Exercises 188–194 are based on the Haywards, California, quadrangle geologic map in the San Francisco folio, No. 193.]

188. An erosion surface occurs between the Jurassic and Cretaceous about one-fourth mile south of the "C" in the word "Concord" on the north margin of the map. Is this an angular unconformity, disconformity, or nonconformity?
189. Is the fault 1.5 miles S50°E of longitude 122°5′W and latitude 37°40′N older or younger than the fault 1.2 miles east of the same reference point? Explain.
190. What is the age of the faulting in the N rectangle? Did it all occur at the same time? Explain the evidence for your answer.
191. What is the name of the structure between the two northwest-trending faults located about 1.5 miles west of the last "n" in the word "Pleasanton" along the east margin of the map?
192. In the NE rectangle, why is the Briones sandstone present east of Cull Creek, but absent to the west?
193. Sketch a generalized cross section along the line A-A from Crow Creek to the northeastern corner of the map. Is the absence of the San Pablo formation in the southwestern portion of the cross section caused by lensing out of the formation as a result of lack of deposition there, or is there an unconformity? In what directions does the unconformity indicated in the legend at the base of the Orinda formation increase in magnitude—northeast or southwest, northwest or southeast? (*Sketch the cross section in the space at the bottom of page 103.*)
194. What structural term is applicable to the body of Jurassic sedimentary rocks just tangent to the 122°5′ meridian at the approximate latitude of 37°32′30″ if bedding is undisturbed? Answer the same question assuming that the bedding is discordant to that of the same formation in nearby areas.

FIG. 7—3 Geologic map of a hypothetic quadrangle for interpretation in Exercises 167—187.

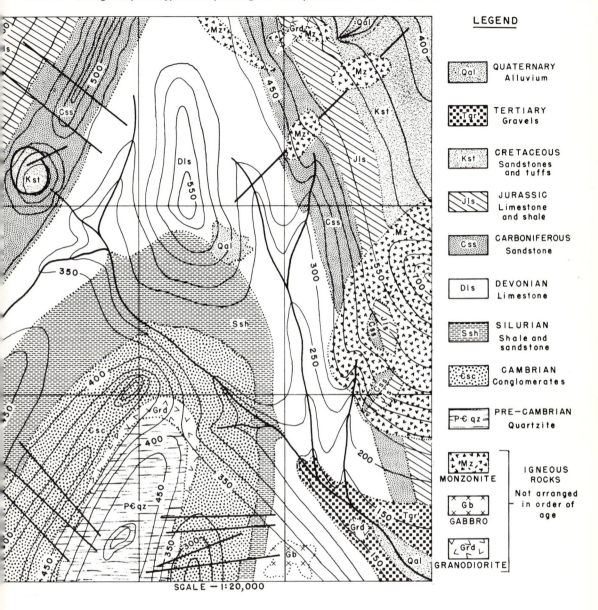

LEGEND

QUATERNARY
Alluvium
Qal

TERTIARY
Gravels
Tgr

CRETACEOUS
Sandstones
and tuffs
Kst

JURASSIC
Limestone
and shale
Jls

CARBONIFEROUS
Sandstone
Css

DEVONIAN
Limestone
Dls

SILURIAN
Shale and
sandstone
Ssh

CAMBRIAN
Conglomerates
Csc

PRE—CAMBRIAN
Quartzite
P∈qz

IGNEOUS
ROCKS
Not arranged
in order of
age

MONZONITE *Mz*

GABBRO *Gb*

GRANODIORITE *Grd*

SCALE — 1:20,000

[Exercises 195–198 are derived from the Silver City, New Mexico, quadrangle geologic map, folio No. 199.]

195. In the Little Burro Mountains area of the W and SW rectangles, what are the relative ages of the faults trending N35°W, N15°E, and N50°E?

196. What is the origin of the relationships of the exposures (areal pattern) of grs, Kb, Kup, and Qgs near the center of the western margin of the quadrangle—T18S, R16W?

197. What name is applied to the zone of northeast-striking faults extending northwest from Silver City—T18S, R14W?

198. What name is applied to the structure bounded by faults and located 3.0 miles south of the northern border of the map and along longitude 108°10'W?

[Exercises 199–204 are based on the Philipsburg, Montana, quadrangle geologic map, folio No. 196.]

199. What are the age relationships of Tagd and Tg in the Mt. Haggin–Clear Creek region of the SE rectangle? Explain the evidence.

200. In the S rectangle, what are the age relationships of Tad, Tgd, and Tagd? Explain the evidence.

201. Locate the short fault trending NW toward the "L" in "Deer Lodge Co." near the eastern border of the E rectangle. According to the legend, what kind of fault is it? Give two other more commonly used names for this type of fault. What is the probable direction of dip of the bs sill, assuming no strike-slip movement? What happened to €f on the northeastern side of the fault?

202. Describe completely in your own words the nature of the structure in the vicinity of Racetrack Peak and Powell Mines southwest to Racetrack Creek. This area is in the southeastern part of the NE rectangle.

203. Find the strike and dip symbol (32°SE) about 1¾ miles south and 3⅜ miles east of the intersection of latitude 46°10'N and longitude 113°20'W. What is the specifically descriptive term (two words) for the structure centered on "Sm" immediately west of the strike and dip symbol?

204. Locate the roughly circular areas of Sm and €rl one mile north and 6½ miles east of the latitude-longitude reference in Exercise 203. What is a good descriptive term (two words) for this sort of stratigraphic development? State whether the most immediate strike and dip symbol indicates the same dip as the relation of contact and contour lines. Draw an enlarged freehand structure section of the area from southeast to northwest crossing both adjacent faults, *in the space below.*

[Exercises 205–212 refer to the geologic maps of the Huntingdon and Hollidaysburg, Pennsylvania, quadrangles, folio No. 227.]

205. What causes the peculiar outcrop pattern of Dcc in the southeastern part of the NE rectangle of the Huntingdon quadrangle?

206. In what direction does the Ore Hill limestone member, €o, of the Gatesburg formation thicken? Consider both maps.

207. What causes the sharp deflection of the outcrop of Oa near a point about 3.8 miles from the southeastern end of cross section E-E′, Hollidaysburg quadrangle?

208. The thrust faults in the southeastern part of the Hollidaysburg quadrangle are of what type?

209. What mistake apparently has been made in the green dip symbols on one of the fold axes near Altoona in the Hollidaysburg quadrangle?

210. Locate a site for a large ganister quarry (see under Tuscarora quartzite, St, in legend). The quarry must be within two miles of an existing railroad, and the beds must dip essentially parallel to the ground surface. Consider both maps.

211. A water well is to be drilled at the road intersection at Foot of Ten (near Duncansville) in the southern part of the NW rectangle of the Hollidaysburg quadrangle. What would probably be the first good aquifer encountered? Is the well likely to be artesian?

212. If the Sinking Valley anticline of the Hollidaysburg quadrangle is the westernmost strong fold of the Valley and Ridge geomorphic or physiographic province, what structural (and topographic feature) is partially shown in the extreme northwestern corner of the quadrangle?

[The following questions refer to the geologic map of the Santa Rita Quadrangle, New Mexico, Map GQ-306.]

213. What is the age of the fault in Martin Canyon (section 10 at bottom of map)? Calculate the heave and throw of the fault, assuming that there has been no strike-slip motion.

214. For section 11 in northern part of map, arrange the following in chronologic order: formations Tg, Po, Ps, Tli, Ml, Klp and the time of faulting. List your evidence.

215. In section 23 identify the relative ages of Ps, Po, Tg, Tli, Klp, and Kp. Cite your evidence.

216. In sections 8 and 9 at the bottom of the map list your evidence for establishing the age sequence among Tba, Tcb, Tbi, Ts, Tk, Tp, Qt, and the faulting.

217. When did faulting occur along the Mimbres fault in the northeastern corner of the quadrangle? What is the evidence?

218. Summarize the record of geologic history contained in the rocks in section 7 in the northeast part of the quadrangle.

219. List in chronologic order the time of formation of the faults and the following rock units in section 21: Po, Klp, Ps, Tw, Tli, Tqm, Dp, Ml, Sofm, Thg, Tg, Qt, Qal, O€b, and Tlq. Identify in section 21 the location of a disconformity, angular unconformity, and nonconformity. Locate in section 21 a sill, dike, and stock.

[Exercises 220–228 are based on the geologic map of the Dromedary Peak Quadrangle, Utah, Map GQ-378.]

220. Near the east border of the quadrangle at latitude 40°35′N, establish the relative ages of €m, €o, €t, Mdg, Mf, gd, and ind, as well as the high-angle faulting. Cite your evidence.

221. When did the thrust faulting occur?

222. What is the nature of the contact between p€bc and q near Superior Peak (111°40′30″W longitude)?

223. The Little Cottonwood stock is cut by several smaller intrusions. What are the relative ages of sd, lq, qm, ld, and py?

224. Why are the high angle faults concentrated around the periphery of the Little Cottonwood quartz monzonite stock? Why do the intermediate dikes occur only at the east end of the quartz monzonite?

225. The long axis of the Little Cottonwood stock trends northeast. Do the margins of the stock probably slope more gently to the NW, NE, or SE? Why?

226. Near the south margin of the map are structures that are almost a cupola and a roof pendant. Where are they? Draw a cross section through them.

227. Where is the Little Cottonwood quartz monzonite most strongly discordant? Where is it most nearly concordant?

228. Calculate the dip of the thrust fault that separates the Mississippian and Cambrian strata at Porcupine Gulch (near SE corner of quadrangle) and 0.5 mile west and 1.0 mile northwest of "P" in "Porcupine."

References

Blyth, F. G. H., 1965, Geological maps and their interpretation: London, Edward Arnold (Publishers) Ltd., 48 p.

Butts, C., 1945, Hollidaysburg-Huntingdon, Pennsylvania Folio: U.S. Geol. Survey, Geol. Atlas of the U.S., folio 227, 20 p.

Calkins, F. C., and Emmons, W. H., 1915, Philipsburg, Montana Folio: U.S. Geol. Survey, Geol. Atlas of the U.S., folio 196, 25 p.

Chalmers, R. M., 1926, Geological maps, the interpretation of structural detail: London, Oxford University Press, 175 p.

Chamberlin, T. C., 1897, The method of multiple working hypotheses: Jour. Geology, v. 5, p. 837–848.

Crittenden, M. D., Jr., 1965, Geology of the Dromedary Peak Quadrangle, Utah: U.S. Geol. Survey, Map GQ-378.

Currier, L. W., 1965, Geologic map portfolio no. 1: A laboratory study of geologic maps and sections: Washington, D.C., Williams and Heinz Map Corp., 64 p. plus 8 maps and cross sections.

Harrison, J. M., 1963, Nature and significance of geologic maps, p. 225–232, in C. C. Albritton, editor: The fabric of geology, Stanford, California, Freeman, Cooper, and Co., 374 p.

Hernon, R. N., and others, 1964, Geology of the Santa Rita Quadrangle, New Mexico: U.S. Geol. Survey, Map GQ-306.

Kupfer, D. H., 1966, Accuracy in geologic maps: Geotimes, v. 10, no. 7, p. 11–14.

Lahee, F. H., 1961, Field geology: New York, McGraw-Hill Book Co., 6th ed., 926 p. (See chapter 20 on interpretation of geologic maps, p. 727–738.)

Lawson, A. C., 1914, San Francisco, California folio: U.S. Geol. Survey, Geol. Atlas of the U.S., folio 193, 24 p.

Mackin, J. H., 1950, The down-structure method of viewing geologic maps: Jour. Geology, v. 58, p. 55–72.

Paige, S., 1916, Silver City, New Mexico folio: U.S. Geol. Survey, Geol. Atlas of the U.S., folio 199, 19 p.

Peterson, D. W., 1960, Geology of the Haunted Canyon Quadrangle, Arizona: U.S. Geol. Survey, Map GQ-128.

Peterson, N. P., 1954, Geology of the Globe Quadrangle, Arizona: U.S. Geol. Survey, Map GQ–41.

CHAPTER 8

Oriented Structures

IN ANY SINGLE REGION certain types of geologic structures can be seen repeated in many different exposures. These structures may be as simple as joint planes or as complex as lineation within the foliation planes of gneiss. The orientation in space of any single structure can be measured, and a compilation of individual orientations can be studied for geometric patterns. Presumably, patterned orientation was caused by a stress field that produced the structures at some time in the past. Theoretical considerations may suggest the orientation of these ancient forces.

Planar Structures

Geologic structures are fractures or patterns in rocks that are larger than the individual mineral grains that comprise the rocks. Most fractures are planar in shape. This is true of nearly all joints and cleavage, as well as many faults. Even curved faults can be considered approximately as a plane in any small area on the curving fault. Most bedding surfaces are planar, with the exception of certain types of cross-beds which are arcuate.

A partial listing of planar geologic structures is as follows:
1. Bedding formed by rocks of contrasting lithology.
2. Color-banding in rocks, such as different-colored layers of sandstone or light and dark layers in gneiss.

3. Tabular igneous intrusions, particularly sills and dikes.
4. Cross-beds in sediments.
5. Joints.
6. Faults.
7. Foliation or schistocity in schist or other metamorphic rocks.
8. Fracture cleavage.
9. Orientation of mica flakes parallel to bedding of sediments.
10. Nearly parallel orientation of tabular crystals of feldspar in gneiss or gneissic granites.
11. Axial plane of a fold.
12. Vein fillings.
13. Microscopic determination of orientation of mica flakes in a schist or phyllite.
14. Orientation of planar structures (such as relic bedding) in xenoliths or roof pendants.

The orientation of any of these planar surfaces can be measured by using a Brunton compass to determine the strike and dip of the plane or by marking the sample so that it can be reorientated in the laboratory for further study.

Lineations

Linear structures are of two types: those forming streaks in the rock, and those consisting of lines formed by the intersection of two planes. Streaks commonly result from sub-parallel orientation of elongate mineral crystals. An example of a family of lines is formed by the trace of bedding on a cleavage surface of slate.

Types of geologic lineations include:

1. Linear orientation of elongate crystals. Figure 8-1 illustrates lineation of hornblende crystals in a gneiss.
2. Slickensides in shear planes.
3. Quartz grains in sandstone have a slight preferred orientation with the c-axis of

FIG. 8–1 Strike and dip of foliation and rake and plunge of lineation in a hornblende gneiss. H—hornblende, Q—quartz, M—muscovite, B—biotite, F—feldspar. The lineation of mineral grains nearly always lies within the foliation plane, if foliation is present.

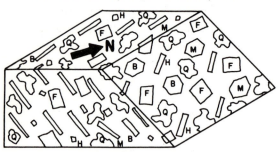

Sketch of Minerals

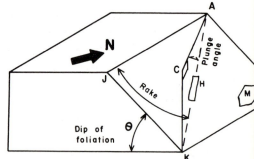

Orientation of lineation and foliation

the quartz tending to be parallel to the flow direction of the water that deposited the grains.

4. Pebbles in conglomerate were deposited with the long axis parallel to the direction of current flow.
5. Long axis of an oolite that has been tectonically deformed from a spherical to elliptical shape.
6. Linear current markings in sediment, such as flute casts, groove casts, and bounce casts.
7. Parallel orientation of high-spired gastropods or straight-shelled cephalopods in a sediment, forming a lineation on the bedding surface.
8. Oriented crests of ripple marks, trending perpendicular to the direction of flow of the current.
9. Orientation of fold axial lines in space (axial line is formed by the intersection of the bedding surface and the axial plane or axial surface).
10. Intersecting line formed by two joints with different orientation.
11. Intersection of two planar faults.
12. Intersection of dike and bedding.
13. Trace of bedding plane on cleavage (or vice versa).
14. Intersection of veins and bedding.
15. Glacial striations.
16. Crest lines of anticlines and trough lines of synclines (the high and low points, respectively, along a folded bedding surface).

The attitude of a lineation is best described by its plunge direction and amount. Plunge direction is the direction toward which the lineation slopes as viewed from above, that is, its orientation in map view. Plunge amount is the angle between the lineation and a horizontal line drawn in a vertical plane that includes the lineation (Fig. 8–1).

If a lineation occurs as a line within a planar structure, the lineation may be described by its rake in the inclined plane. *Rake* is the angle between the lineation and a horizontal line, both lying in the plane of the planar structure (Fig. 8–1). Rake and plunge amount become identical values as the planar structure approaches a vertical attitude.

Intersection of Two Planes / Two planes intersect along a straight line. The orientation of this line can be calculated if the attitude of each plane is known. Formal descriptive geometry solutions to this problem are given by Nevin (1949, p. 372–374), Billings (1954, p. 459–463), and Lahee (1961, p. 740–742).

A quick and accurate solution is obtained from a scale drawing of a simple structural contour map of the two intersecting planes. *A typical problem is solved in Figure 8–2. There a vein has an attitude of S75°W 30°SE, and a fault has an attitude of S10°E 40°SW. It is desired to find the plunge of the line formed by the intersection of the two planar structures. First, draw a contour (strike) line on the vein (AB) and on the fault (BF). These two contours have the same elevation and intersect at the* common point B. If the contour interval is of unit value, the horizontal distance (AC) between parallel contours AB and CG on the vein is plotted equal to the cotangent of the dip angle (30°) of the vein. In like manner, the horizontal distance (EF) between GE and BF equals the cotangent of the dip (40°) of the fault. Contours CG and EG are the same elevation, and point G is common to both contours. Because points G and B are common to both the fault and vein, the line BG is the direction of the line of intersection of the fault and vein. The length of BG equals the cotangent of the plunge angle of the line of intersection.

Once the plunge direction is determined by this graphical method, the plunge angle can be computed from the attitude of one of the planes, by either of the following formulae:

Tan plunge angle = (Tan true dip angle) (Cos angle between true dip and plunge directions). (58)

Tan plunge angle = (Tan true dip angle) (Sin angle between strike and plunge directions). (59)

Plunge is actually an apparent dip of the planar structure, and Formula 58 is a modification of Formula 1 in Chapter 1, which was developed to calculate apparent dip from true dip.

The following derivation yields a formula for converting plunge to rake (reference Fig. 8–3).

BAE = Rake.

$$\text{Cos BAE} = \frac{AB}{AE} = \frac{AC \text{ Cos BAC}}{AC \text{ Sec CAE}} = (\text{Cos BAC})(\text{Cos CAE}).$$

Cos Rake = (Cos angle between plunge direction and strike of plane) (Cos plunge angle). (60)

Orientation Plots on Equal-area Projection

The attitude of a lineation or a plane is best described in a spherical coordinate system. An equal-area projection of a spherical grid system is the customary way to portray three-dimensional orientations on a planar plot.

Plotting Planes and Lines in a Spherical Space / Planar structures or lineations can have any conceivable orientation in space. Numerous measurements of these structures are commonly plotted onto a map projection of a sphere in order to search for a preferred orientation.

A lineation can be plotted by its plunge direction and amount, assuming that the line passes through the center of the sphere and that each end of the line pierces the surface of the sphere. Such a plot is symmetrical about the center of the sphere, with the plot on the upper half of the sphere being the mirror image of the lower piercing point. Consequently, it is customary to use only the lower hemisphere plot for most studies. (Some linear orientations not only are a line in space, but also have a vectorial sense,

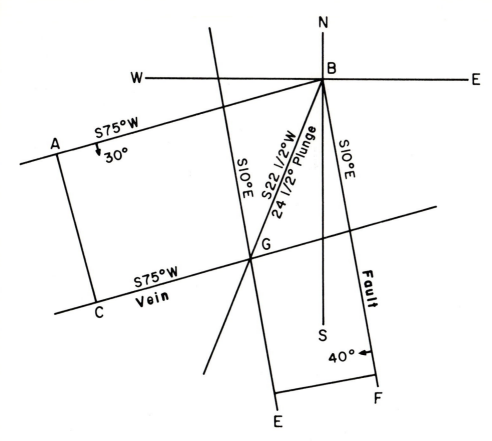

FIG. 8–2 Finding the attitude of a line formed by the intersection of two planes.

FIG. 8–3 Diagram for derivation of Formula 60, showing relation between rake (angle BAE) and plunge (angle CAE).

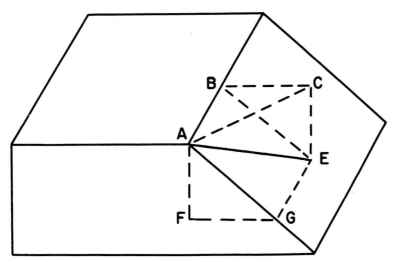

such as the direction of flow of water indicated by the lineation markings in a flute cast.) Figure 8–4 is a plot of a lineation plunging 30°N45°E, showing the lineation as a point at A where the line pierces the sphere. The peripheral circle representing the horizontal surface of the top of the hemisphere is known as the *primitive circle*.

Similarly, a plane can be represented in a spherical space. Figure 8–5 is a plot of a plane dipping 50°S62°W. The outline HJK is a plot of the plane. The strike of the plane (HK) passes through the center of the sphere, and arc KJH is the intersection trace of the plane with the surface of the sphere.

This same plane can be represented on the sphere by a single point at position A in Figure 8–6. Imagine the plane intersected by a line normal to that plane, placed with the intersection at the center of the sphere. Point A is the plot on the lower hemisphere of the position where the normal line intersects the sphere. This procedure is most useful for plotting data obtained in petrofabric studies using a microscope with universal stage.

The same plane may be plotted a third unique way on the same spherical projection. Planar orientation data generally are recorded in the field as strike and dip of the plane. The dip line is the line drawn at right angles to strike in the planar surface. The piercing point of this dip line can be plotted directly on the sphere, without calculating the orientation of a line normal to the planar structure. (See the same plane plotted as point B in Figure 8–6.) Direct plots of the dip line may save time in studies of joints, cross-beds, cleavage, or field measurements of foliation.

Figures 8-4 through 8–6 are simple representations of a three-dimensional space. It

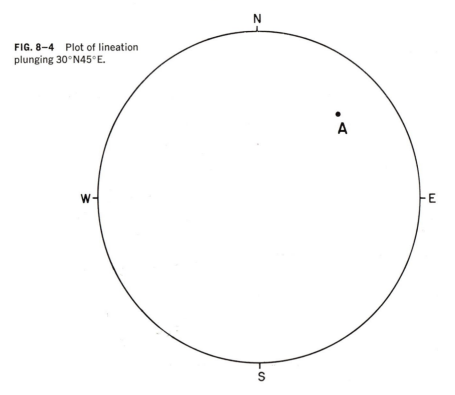

FIG. 8–4 Plot of lineation plunging 30°N45°E.

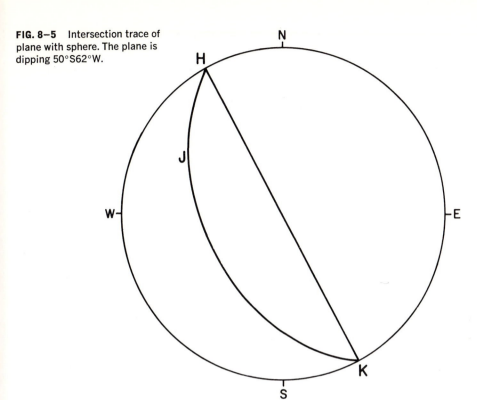

FIG. 8–5 Intersection trace of plane with sphere. The plane is dipping 50°S62°W.

FIG. 8–6 Point A is the plot of a line normal to a plane dipping 50°S62°W. Point B is the piercing point of the dip line of the plane. Dashed line shows intersection of the plane and hemisphere.

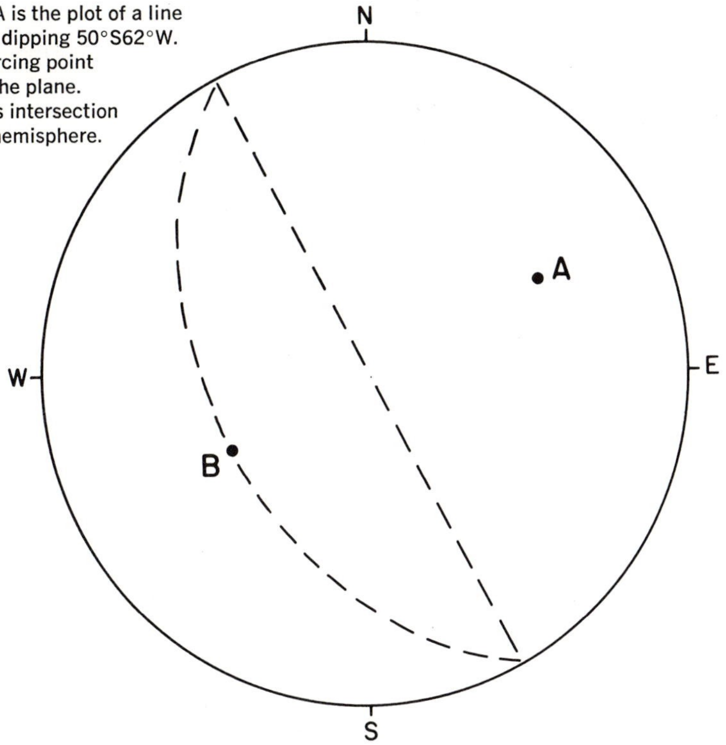

is necessary to map orientations accurately on the spherical plot. Several types of map projections can be used to plot the piercing point or intersection trace of a plane. For orientation studies in structural geology it is customary to use an equal-area projection. Equal-area projections have the useful property that an area on the surface of a sphere is represented by the same area on the projection, no matter where it is located. For example, a piece of spherical surface equal in area to 2% of the total surface of the sphere has the same size on an equal-area net no matter whether it is at the center or anywhere on the periphery of the plot. Similarly, if plots of structural geology piercing points on the equal-area net appear to have twice the density (abundance) in one part of the net compared to another part, the density truly is twice as great when the points actually are plotted on a three-dimensional sphere. In other words, apparent concentrations on an equal-area plot are real concentrations.

Two types of equal-area projections of a sphere can be used. Both were developed by a mathematician named Lambert late in the eighteenth century. One is a projection centered on the south pole of a global pattern of latitude and longitude and is called the Lambert azimuthal equal-area projection (Fig. 8–7). The other type is centered on the equator and a central or "prime" meridian (Fig. 8–8) and is known as a Lambert meridional equal-area projection. Each type is preferred for certain structural geology uses, so both are treated briefly in this chapter.

The difference between the two kinds of equal-area projections depends on the choice of orientation of a reference coordinate system. No matter which projection is used, a point on the surface of a sphere will be at the same position relative to the north horizontal direction plot on the periphery of the projection. Only the plot of the reference system of coordinates changes relative to north on the periphery; the plot of the piercing point remains the same.

Lambert Azimuthal Equal-area Projection / In this projection azimuths are readily plotted as an angle measured clockwise from north, plotted about the observer's nadir (the point directly beneath the observer), that is, about the center of the projection net. Data can be plotted as follows, using Figure 8–7 for a coordinate system.

1. Plotting a lineation. A line is plotted by its plunge direction and amount. Plunge direction is read on the periphery scale. Plunge amount is read by counting degrees in toward the nadir from the periphery of the plot. (Figure 8–7 is plotted into 10° circles for plotting plunge amount; for more accurate plotting you may wish to use a drafting compass with pencil to interpolate linearly the position of circles at 2° increments.) Figure 8–9 is a plot of a line plunging 30°N45°E, using this projection base. (Compare with Figure 8–4.)

2. Plotting a plane.

   A. NORMAL POLE METHOD. The most common procedure to indicate orientation of a plane is to plot the position of a line normal to the plane. If a plane dips 50°S62°W, its normal line plunges 40°N62°E. Draw a radius of the large circle pointing in the plunge direction of a line normal to the plane. The piercing position of the normal line is plotted by counting inward from the periphery along that radius a number of degrees equal to the plunge amount of the line normal to the plane. Figure 8–10 is a normal pole plot of the plane dipping 50°N62°W.

FIG. 8–7   Lambert azimuthal equal-area projection, with grid spacing in 10° increments.

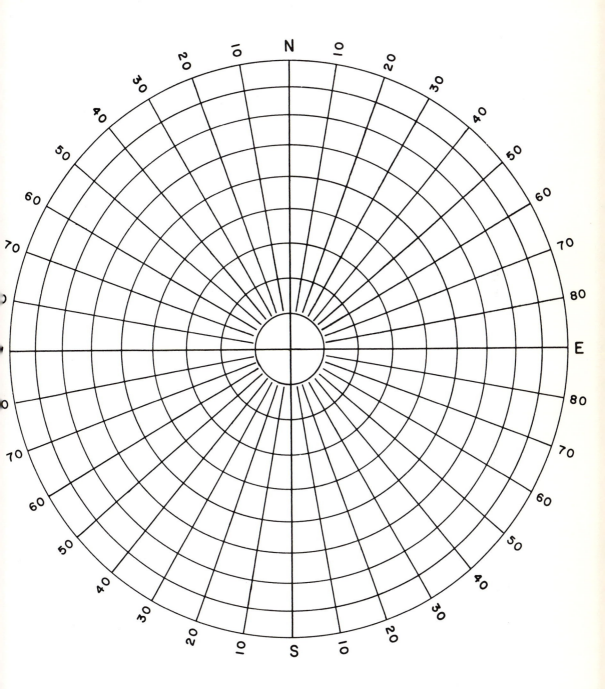

FIG. 8—8 Lambert meridional equal-area projection (Schmidt net). (Courtesy of the Johns Hopkins University.)

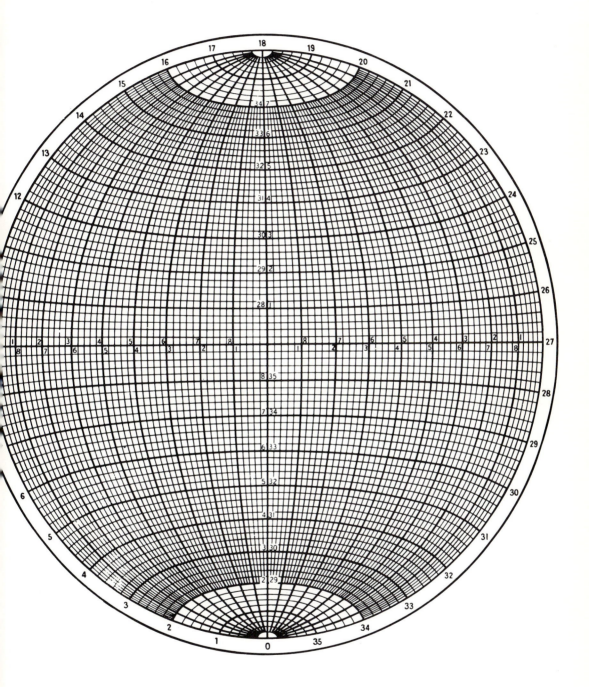

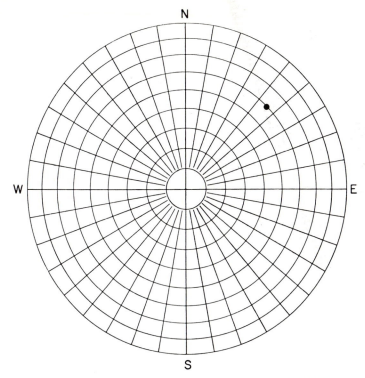

FIG. 8-9 Lambert azimuthal equal-area projection plot of a line plunging 30°N45°E.

FIG. 8-10 Normal pole plot of a plan dipping 50°S62°W, using a Lambert azimuthal equal-area projection.

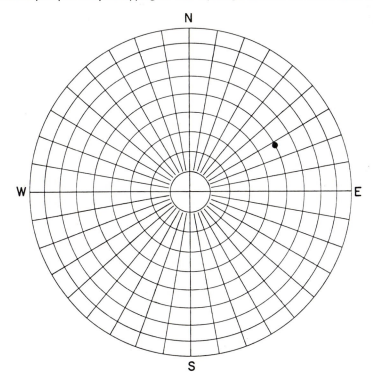

B. Dip line method. Draw a radius line in the dip direction. Count in from the primitive circle along that line the number of degrees equal to the dip amount. A point at that position is the plot of the dip line piercing the sphere. Figure 8-11 shows a dip line plot of a plane dipping 50°N62°W, using an azimuthal projection.

Instead of marking directly on Figure 8–7 to plot piercing points, cover it with a protective overlay of thin paper so that the underlying coordinate system may be used again. Draw a peripheral circle on the overlay sheet, and mark it with a north compass direction.

Billings (1954, p. 108–115) describes an example of plotting joints, using an azimuthal equal area projection. A normal line should be plotted when studying steeply dipping joints, so that differences in strike direction are most apparent. Sheet jointing with low dips nearly parallel to the ground surface is most vividly plotted using the dip line method. Orientation studies of cross-beds in nearly horizontal sediments are done best using plots of the dip line, so that in three-dimensional orientation the dip direction of the cross-bed is most apparent, yet the variation in dip amount is obvious.

Lambert Meridional Equal-area Projection / In addition to plotting piercing points of lines, the meridional projection allows one to plot the intersection trace of a plane onto the spherical surface. The projection net of Figure 8–8 is sometimes called a Schmidt net, after the German who first called attention to its applications in structural petrology (1925).

An overlay sheet must be used, so that the plotted points can be rotated about the center of the projection. Mount Figure 8–8 on thin cardboard to make a plotting instrument. Pierce the center of the projection mounted on thin cardboard and the overlay sheet with a thumb tack to serve as an axis of rotation. Scotch or other clear plastic tape placed over the backside center of the overlay sheet prevents enlargement of the hole when the overlay sheet is rotated. Draw a primitive circle on the overlay sheet, and mark the north direction. Directions on a meridional net customarily are measured clockwise from south, because of convenience in working with a universal stage microscope.

1. Plotting a lineation.

A line pierces the sphere at the same point as in an azimuthal projection. However, the plotting procedure is a bit different because of the meridional coordinate system. Take the example of a line plunging 30°N45°E. Draw the primitive circle and label north on the overlay sheet. Mark the plunge direction (N45°E) on the primitive circle, then rotate until the mark is in an east-west position. Count in 30° along the east-west line from the periphery and plot a point. Then rotate back until the reference north line of the overlay coincides with north on the meridional underlay sheet. The point on the overlay sheet now is located correctly (see Fig. 8–12), in the same position relative to the north mark on the peripheral circle as shown in Figures 8–4 and 8–9.

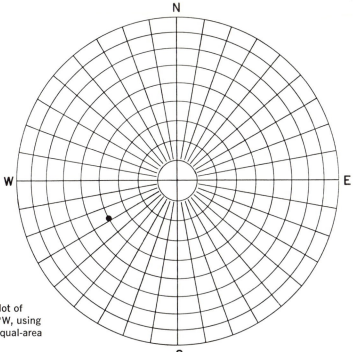

FIG. 8–11 Dip line plot of plane dipping 50°S62°W, using a Lambert azimuthal equal-area projection.

FIG. 8–12 Lambert meridional equal-area projection plot of a line plunging 30°N45°E.

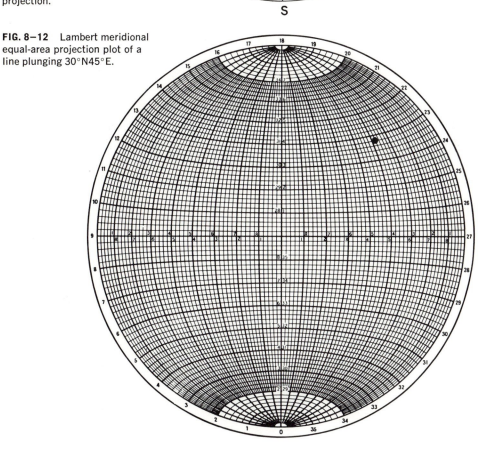

2. Plotting a plane.

Consider the example of a plane dipping 50°S62°W.

A. NORMAL POLE METHOD. The pole normal to the plane plunges 40°N62°E. Plot this pole according to the piercing point of the line, using the rotation procedure for the overlay as described for lineation plots with the meridional projection. The plot of the normal pole is point A in Figure 8–13.

B. DIP LINE METHOD. The dip line of the plane plunges 50°S62°W. Plot this dip line just as if it were a lineation. Point B in Figure 8–13 illustrates the correct plot.

C. PLANAR TRACE METHOD. The meridional projection is ideally suited for plotting the intersection trace of the plane with the sphere. The plane strikes N28°W. On the overlay draw a diameter of the primitive circle trending N28°W. Mark a dip direction radius oriented S62°W. Rotate the overlay until the S62°W line lies above an east-west line on the underlay sheet. Count in 50° from the primitive circle to get the plot of a dip line piercing point. On the overlay sheet trace the meridional great circle 50° from the periphery, intersecting the strike line at the north and south marks on the underlay sheet primitive circle. Rotate the overlay sheet back to the starting position, with north coinciding on the overlay and underlay sheets. The trace of the plane on the sphere is now properly oriented (arc and strike line in Fig. 8–13).

Detailed use of a meridional equal-area net in structural petrology is described by Knopf and Ingerson (1938), by Turner and Weiss (1963), and by Whitten (1966). The last two references also consider its use in other aspects of structural geology.

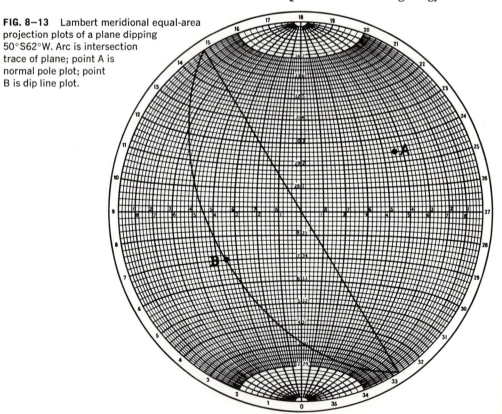

FIG. 8–13 Lambert meridional equal-area projection plots of a plane dipping 50°S62°W. Arc is intersection trace of plane; point A is normal pole plot; point B is dip line plot.

FIG. 8–14 Grid for use with counters from Figure 8–15 to determine concentration of points in equal-area plot.

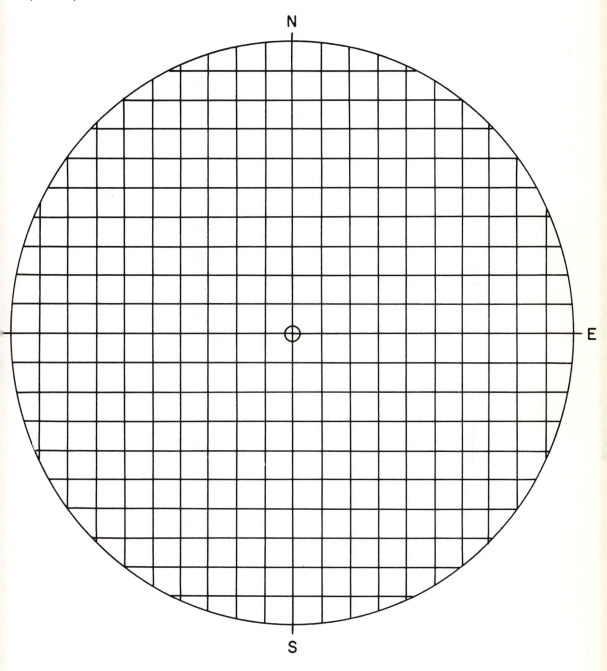

FIG. 8—15 Counters for determining concentration of points in equal-area plot.

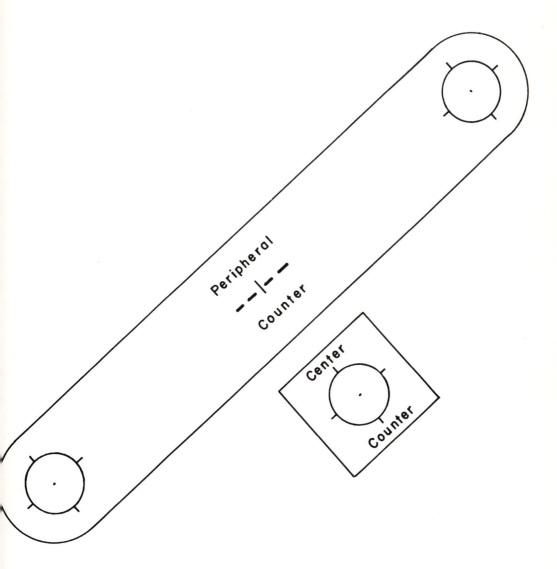

Counting and Contouring Orientation Diagram / An equal-area projection showing piercing points of lines is called a *point diagram*. It is customary to express point concentrations as percent of total points that occupy any specified 1% area of the hemisphere. This is accomplished in the following manner.

The overlay plot of points prepared from Figure 8–7 or 8–8 is placed on the grid of Figure 8–14, with the center of the overlay at the center of the grid and with the north-south direction parallel to the vertical grid lines. On the overlay sheet place a dot at each grid intersection.

Using a razor blade, cut out the center counter and peripheral counter in Figure 8–15 along their outer edges, and remove the circles centered inside the cross-lines of the counters. Each small circle has an area equal to 1% of the large plotting net. Cut out the dashed line of the peripheral counter, making two parallel slices with the blade.

Place the center counter at each grid intersection occurring far enough inside the peripheral circle of Figure 8–14 that the periphery of the net is not visible through the center counter. Label each grid intersection with the number of points visible inside the center counter. Figure 8–16 illustrates this procedure.

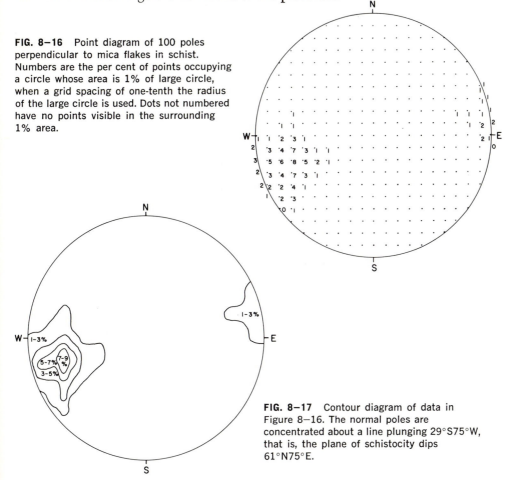

FIG. 8–16 Point diagram of 100 poles perpendicular to mica flakes in schist. Numbers are the per cent of points occupying a circle whose area is 1% of large circle, when a grid spacing of one-tenth the radius of the large circle is used. Dots not numbered have no points visible in the surrounding 1% area.

FIG. 8–17 Contour diagram of data in Figure 8–16. The normal poles are concentrated about a line plunging 29°S75°W, that is, the plane of schistocity dips 61°N75°E.

If the peripheral circle of Figure 8–14 is visible in the center counter, then the intersection of the grid lines located close to the margin should be studied using the peripheral counter. Place the slot of the counter over a thumb tack pierced through the center of Figure 8–14 and place the center of a circle of the peripheral counter over the grid intersection. At the grid intersection dot on the overlay sheet, indicate the total number of points visible in *both* windows of the peripheral counter. Points within 1% of the total area of the large circle are still counted when both windows are used; part of each window of the peripheral counter lies outside the domain of the equal-area plot. Count along the periphery, moving clockwise from north until you have counted points centered about all grid intersections that are located close enough to the margins of Figure 8–14 that the peripheral circle is visible. Then proceed around the peripheral circle, this time counting points in both openings of the peripheral counter when the counter is centered at the intersection of each grid line with the periphery. Figure 8–15 is a completed point diagram.

The numbered grid of points can be marked off into fields with different concentrations of points, much as relief maps sometimes are shaded different colors to delimit regions within a certain range of elevation. Figure 8–17 is a *contour diagram* of the points in Figure 8–16, placing boundary limits at 2% increments, that is, at 1%, 3%, 5%, and 7%. For clarity and ease of interpretation no more than six boundary isopleths should be used. Ideally, the contour interval should be uniform. When a contour intersects the primitive circle of the overlay sheet at one place, the contour should reappear at the opposite side of the diagram; all contours must be closed contours. Figure 8–18 illustrates two examples of contour diagrams resulting from structural analysis.

Slightly different procedures for preparing contour diagrams are described by Knopf and Ingerson (1938, p. 245–251), Fairbairn (1949, p. 285–291), and Turner and Weiss (1963, p. 61–64). All of the methods produce very similar contour diagrams.

FIG. 8–18 Petrofabric diagrams from near Turoka, Kenya. Contours at 9%, 7% 5%, 3% and 1% per 1% area. A. 284 lineations. B. 429 foliation poles, showing girdle. (From *Structural Analysis of Metamorphic Tectonites* by F. J. Turner and L. E. Weiss. © 1963, the McGraw-Hill Company, Inc.)

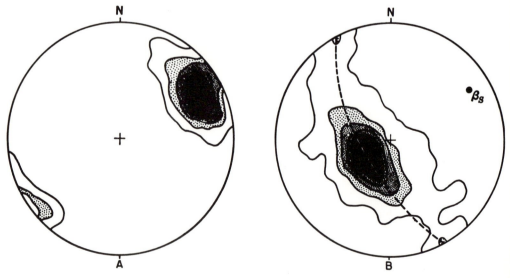

Girdles / The meridional equal-area projection enables a geologist to measure the orientation of a great circle streak of data points that occur as a *girdle* across the projection sphere. Such a girdle of data points is much like the stars that form a great circle Milky Way band across the sky. Figure 8–19 shows a plot of 20 points that lie along a girdle. In order to measure the attitude of the girdle plane, place the overlay sheet centered on a meridional equal-area net, and then rotate the overlay sheet until the points seem to lie along one of the meridians of the projection. When the best fit is obtained, draw on the overlay sheet a best-fit line tracing a meridional path extending from north to south pole of the underlay sheet. Draw a straight line connecting the ends of the meridional line and passing through the center of the plot; this is the strike of the girdle. Dip amount of the girdle plane is measured inward from the periphery of the meridional net. Figure 8–20 shows the girdle in this position. Next rotate the overlay sheet back to a north-south orientation, and read the strike direction of the girdle from the overlay sheet. Dip direction is, of course, perpendicular to strike. Figure 8–21 shows the best-fit girdle through the points in Figure 8–19, a plane dipping 25°N67°E.

Structural petrologists like to use a meridional equal-area projection, because it permits them to search for girdles. The procedure for plotting points on a meridional projection is more cumbersome than for an azimuthal projection. A piercing line penetrates the sphere at the same point relative to north on the primitive circle, regardless of an azimuthal or meridional projection, so it is recommended that points be plotted using an azimuthal net (Fig. 8–7), and after contouring, then search for a girdle pattern by transferring the overlay sheet onto a meridional net (Fig. 8–8). This procedure permits maximum speed with flexibility of analysis.

Figure 8–18, part B illustrates a girdle pattern oriented N30°W 65°S60°W with a concentration of points along a line plunging 65°S55°W.

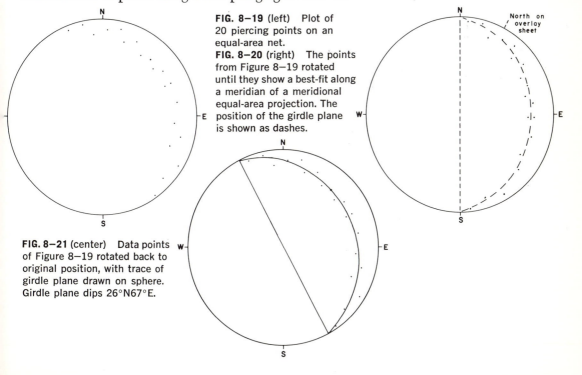

FIG. 8–19 (left) Plot of 20 piercing points on an equal-area net.

FIG. 8–20 (right) The points from Figure 8–19 rotated until they show a best-fit along a meridian of a meridional equal-area projection. The position of the girdle plane is shown as dashes.

FIG. 8–21 (center) Data points of Figure 8–19 rotated back to original position, with trace of girdle plane drawn on sphere. Girdle plane dips 26°N67°E.

Direction Orientation Diagrams

For certain geological purposes only the direction of a planar structure is recorded. Examples include direction of cross-bed inclination, in situations where the master bedding is horizontal; presumably the transporting current flowed toward the direction of dip of the cross-bed. Joints visible on air photos often can be measured only according to their trend (strike direction) of lineaments; strike of joints, rather than dip, is plotted in this situation.

Lineation directions also can be plotted according to orientation in plan view. Common examples include crest line of oscillation ripples, groove casts, flute casts, and the long axis of straight-shelled cephalopods. The crest line of ripples is oriented perpendicular to current movement. The other lineations tend to be parallel to current flow. The tapered end of a straight-shelled cephalopod tends to become oriented pointing upcurrent. The sharp end of a flute cast is the upcurrent direction. A groove cast is parallel to the current, but the upcurrent end usually cannot be identified. In situations where the upcurrent direction of a lineation is indistinguishable (symmetrical ripple marks and groove casts, of the situations listed), current direction should be plotted through a 180° range, because the other half of the compass rose would contain the direction of the other end of the lineation.

Potter and Pettijohn (1963) have written a book devoted to analysis of paleocurrent direction in sediments.

Figure 8–22 is a hypothetical histogram of cross-bed orientation data. In a histogram the *area* of a given class is proportional to the percentage of total values in that class. For a histogram plotted on linear axes, plotted height of a class is proportional to percentage abundance in the class. A linear abscissa, such as the one in Figure 8–22, may be misleading. Actually the cross-bed distribution is unimodal with a mean direction of approximately N20°E, even though the histogram appears bimodal at first glance.

A circular histogram is a plot of class frequency on a compass rose. The circular histogram of Figure 8–23 shows the same data as Figure 8–22. Direction is plotted on a compass rose; relative abundance is plotted with radius equal to height of class in Figure 8–22. This procedure is used by most geologists to portray distribution of directional data (Potter and Pettijohn, 1963, p. 263).

The procedure of Figure 8–23 violates one concept of histogram theory. The area plotted for a class is not proportional to class abundance. The area of a segment of a circle is given by:

$$\text{Area} = \frac{C}{360} \pi r^2. \tag{61}$$

where C equals degrees arc of segment. Radius is proportional to the square root of number of individuals in a class. Consequently, in order to have proportional area plots on a circular histogram, the radius of a 20% abundance class should be $\sqrt{2}$ times greater than the radius of a 10% abundance class. Figure 8–24 is an equal-area circular histogram plot of the data from Figure 8–22. The mental impression gained

FIG. 8–22 Histogram of cross-bed orientation data, with mean direction of N21°E.

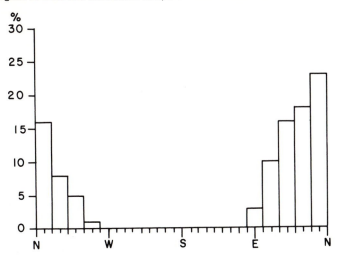

FIG. 8–23 Circular histogram using class radius equal to class height in Figure 8–22.

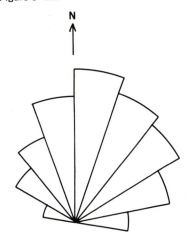

FIG. 8–24 Equal-area circular histogram of data from Figure 8–22.

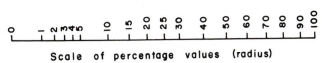

Scale of percentage values (radius)

from glancing at Figure 8–24 gives a truer representation of percentage abundance of classes than does Figure 8–23.

In these considerations of determining current direction from cross-bedding, it is assumed that the master bedding is horizontal and that the original dip of the cross-beds has not been tilted by subsequent deformation of strata. The problem of removing the effect of post-depositional tilting of cross-beds is considered in Chapter 11.

If the bedding is horizontal, current lineations are measured by plotting the azimuth of a line viewed in plan view. For determining the original direction of lineation on a tilted bedding surface, record the rake of the lineation relative to strike of the bedding (Fig. 8–25, part a). Rotation of the bedding back to horizontal will not alter the angle *measured on the bedding surface* between the lineation and the original strike line drawn on the bedding. In Figure 8–25, part b, the current direction can be calculated by subtraction of the rake angle from the azimuth of the strike line.

FIG. 8–25 Determining original direction of current lineation on a bedding surface tilted after deposition. View *a* is straight-shelled cephalopod in present situation. View *b* shows bedding surface rotated about strike line to horizontal position, with current flowing toward S35°W.

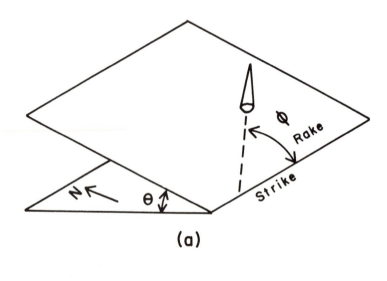

(a)

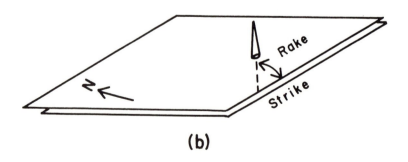

(b)

Preferred Orientations

If directions on an orientation diagram are tightly clustered within a small area, obviously a patterned or preferred orientation is present. One is more confident of a preferred orientation if many data points are plotted, rather than just a few. When the data are not tightly clustered, then the decision of presence or absence of preferred orientation is more difficult. This situation is ideal for the proper application of statistical analysis. Statistical procedures are unnecessary for detecting "obvious" preferred orientation. Statistical techniques do allow one to make objective decisions about presence or absence of preferred orientation.

Analysis of azimuth directions is better understood than proper procedures for studying orientation on a spherical surface. Interpretation of azimuth orientation of sedimentary structures is superbly treated by Potter and Pettijohn (1963). Trigonometric calculation of a vector mean direction is described by Curray (1956), who also presents a statistical test to compare a preferred orientation with a uniform orientation. Durand and Greenwood (1958) evaluate Curray's procedure in light of statistical theory and suggest slight modifications and improvements. Raup and Miesch (1957) and Dennison (1962) consider the problem of an adequate sample size for a desired level of confidence in statements about the mean direction of azimuth data.

Three-dimensional orientation data are more complex to analyze. Flinn (1958) considers the problem of detecting a significant orientation of piercing points of lines on a map of a sphere. He concludes that most workers have compared their data with a uniform distribution of points, and that they have claimed a preferred orientation if their data differ with statistical significance from a uniform or homogeneous distribution. A more appropriate mathematical model is to compare orientation data with points randomly distributed on the surface of a sphere. If the distribution of data points on areas of a spherical surface does not vary significantly from the clustering of points into areas as a result of a random distribution, then a preferred orientation is not present.

Several other references can be cited as useful in describing techniques for detecting preferred orientation in a spherical space. Krumbein (1939) described methods for measuring and detecting preferred orientation in pebbles; the same techniques can be applied to other fabric analyses. Spencer (1959, p. 478–480) worked with the problem of detecting significant preferred orientation of joints. Cox and Doell (1960, p. 668–673) present formulae for calculating confidence limits on the reliability of a mean direction of points which are concentrated in a single area on the spherical surface.

Exercises /

229. The attitudes of 100 joints were measured in an area of granite exposures 150 by 300 feet in size. Prepare a normal pole contour diagram showing the percentage distribution of the attitudes of the joints at that locality. Then formulate a hypothesis for the origin of the different joint sets.

[*Strike and dip*]

1. N3°E 8°NW
2. N75°E 89°NW
3. N10°E 81°SE
4. N73°W 70°NE
5. N40°E 45°NW
6. N76°W 90°
7. N65°W 75°NE
8. N10°W 15°SW
9. N7°E 77°SE
10. N71°W 89°NE
11. N89°W 86°NE
12. N68°W 83°SW
13. N15°E 72°SE
14. N61°W 78°SW
15. N70°W 88°SW

[*Dip*]

16. 87°S63°E
17. 78°N32°E
18. 89°S2°E
19. 89°S80°E
20. 86°N17°E
21. 17°S65°W
22. 87°N76°W
23. 86°N19°E
24. 17°S78°W

25. 12°S86°W
26. 89°S17°W
27. 85°S15°W
28. 77°S43°W
29. 70°S75°E
30. 85°S3°W
31. 85°S13°W
32. 10°S70°W
33. 75°N81°W
34. 10°S80°W
35. 75°N81°W
36. 45°S72°E
37. 63°S72°E
38. 88°N63°E
39. 71°S32°E
40. 29°N63°E
41. 13°S75°W
42. 70°N13°E
43. 72°S73°E
44. 87°N12°E
45. 15°N88°W
46. 87°S82°E
47. 88°N15°E
48. 89°N17°W
49. 73°S79°E
50. 69°N80°W

51. 50°S20°E
52. 4°S61°W
53. 18°S55°W
54. 87°N24°E
55. 77°S77°E
56. 88°S10°W
57. 65°N77°W
58. 74°N24°W
59. 85°N82°W
60. 75°N72°W
61. 78°due S
62. 4°N60°E
63. 14°N74°W
64. 23°N67°W
65. 67°N84°W
66. 68°N76°W
67. 74°N27°E
68. 81°S73°E
69. 67°N75°E
70. 67°N71°W
71. 12°N85°W
72. 83°S8°W
73. 67°N37°E
74. 70°S83°E
75. 62°N27°E
76. 75°S17°W

77. 80°N74°W
78. Vertical
    Strike N14°E
79. 84°S83°E
80. 56°S5°W
81. 3°S55°W
82. 89°N72°W
83. 76°N12°E
84. 84°S77°E
85. 84°N38°E
86. 78°S17°W
87. 73°N75°W
88. 76°S80°E
89. 72°N78°W
90. 10°N60°W
91. 82°S1°E
92. 76°N75°W
93. 87°S72°E
94. 78°N15°E
95. 66°N74°W
96. 13°N42°W
97. 22°N55°W
98. 14°N75°W
99. 17°N65°W
100. 2°N30°W

230. The following planar orientations are joints from a 2-acre exposure of a gabbro intrusive body. Prepare a normal pole contour diagram showing the percentage distribution of the attitudes of joints at that locality. Describe the joint system, and postulate mechanisms that might have produced joint patterns.

[*Strike and dip*]

1. N74°W 10°NE
2. N10°W 81°NE
3. N12°W 89°SW
4. N10°W 79°NE
5. N45°E 89°NW
6. N3°W 84°NE
7. N43°E 87°NW
8. N38°E 88°SE
9. N50°E 86°SE
10. N40°E 75°SE
11. N75°W 3°NE
12. N59°E 40°SE
13. N39°W 72°NE
14. N58°E 69°NW
15. N85°W 3°NE

[*Dip*]

16. 89°S89°W
17. 75°S46°E
18. 48°N46°W
19. 67°N36°E

20. 5°N17°E
21. 6°N19°E
22. 85°N46°W
23. 90°N75°E
24. 1°S35°W
25. 2°N12°E
26. 89°N81°E
27. 63°N17°W
28. 88°N42°W
29. 42°S42°W
30. 5°N22°E
31. 89°S45°E
32. 85°S47°E
33. 53°N16°E
34. 85°S85°W
35. 90°N45°W
36. 65°N63°E
37. 14°N1°W
38. 3°N25°E
39. 87°S83°W
40. 74°S63°W

41. 87°N41°W
42. 80°N49°W
43. 83°N46°W
44. 78°N85°E
45. 85°S88°W
46. 87°E
47. 4°N5°E
48. 2°N6°E
49. 83°N45°W
50. 86°S72°W
51. 70°S80°E
52. 66°N86°E
53. 5°S5°E
54. 80°S77°W
55. 60° due W
56. 75°S42°E
57. 77°N87°E
58. 45°S23°W
59. 78°S81°E
60. 2°S42°W
61. 83°S42°W

62. 83°S80°W
63. 74°N84°E
64. 80°S45°E
65. 77°N40°W
66. 75°S85°W
67. 82°S42°E
68. 84°S47°E
69. 72°S82°W
70. 77°N43°W
71. 2°S85°E
72. 74°S2°W
73. 37°S65°E
74. Horizontal
75. 74°S78°W
76. 78°N47°W
77. 76°S81°W
78. 84°N83°E
79. 76°N82°E
80. 80°N43°W
81. 86°S43°E
82. 81°S84°W

83. 89°S82°W	88. 86°S27°E	93. 67°S80°W	98. 89°S88°E
84. 84°N73°W	89. 70°N50°W	94. 84°N53°W	99. 88°N51°W
85. 28°N45°E	90. 75°N75°W	95. 80°S51°E	100. 60°S20°E
86. 18°S60°W	91. 68°N39°W	96. 85°N78°E	
87. 84°N78°W	92. 76°S79°W	97. 84°N38°W	

231. The Tensleep Sandstone is a very thickly bedded, cross-bedded sandstone. An exposure can be reached by climbing the walls of a canyon, and at close range master-bedding is difficult to distinguish from cross-bedding. The data below are dips of planar surfaces, most of which are cross-beds, but some are master-bedding planes. The master beds dip 28°N80°E. Plot all the planes, using the dip line method. Eliminate the eight points closest to the master-bedding plotted position, because within the limits of accuracy of the field work some master-beds are measured, with recorded values slightly different from the average dip of the master-bedding. Prepare a contour diagram of the 50 points representing cross-beds. What was the approximate average dip and strike of the cross-beds at the time of deposition, that is, before tilting of the master-bedding by orogeny?

1. 5°N60°W	16. 15°N65°E	31. 3°N54°E	45. 5°S78°E
2. 1°S85°W	17. 35°S45°E	32. 6°S45°W	46. 22°N85°E
3. 3°N57°E	18. 20°N55°E	33. 8°N28°E	47. 12°S65°E
4. 10°S55°W	19. 23°S78°E	34. 13°N68°E	48. 2°N78°E
5. 7°N85°W	20. 6°S79°E	35. 9°N47°E	49. 4°N36°E
6. 15°S76°E	21. 8°N65°E	36. 3°S45°E	50. 6°N85°E
7. 8°N36°E	22. 5°S54°W	37. 13°N43°E	51. 5°S10°W
8. 20°N55°E	23. 10°S5°W	38. 4°S65°W	52. 3°N85°W
9. 17°N80°E	24. 3°N35°W	39. 2°S48°W	53. 16°N67°E
10. 28°N80°E	25. 45°S65°E	40. 5°N63°E	54. 13°S78°E
11. 10° due east	26. 29°N75°E	41. 3°N74°W	55. 27°N87°E
12. 19°N88°E	27. 5°N16°E	42. 3°N49°E	56. 17°N78°E
13. 26°N76°E	28. 27°N74°E	43. 17°N86°E	57. 1°S50°W
14. 5°S23°E	29. 8°N68°E	44. 10°N46°E	58. 1°N89°E
15. 26°N5°E	30. 15°N60°E		

232. The data consist of lineation directions in a metamorphic rock, obtained by universal stage determination of the orientation of the C-axis direction of pyroxene crystals. Prepare a contour diagram, and describe the resulting preferred orientation.

1. 30°N80°W	15. 36°S38°W	29. 2°S58°E	43. 52°S3°W
2. 46°S56°W	16. 56°S8°W	30. 30°S36°E	44. 10°S48°E
3. 8°N80°W	17. 58°S47°W	31. 55°S65°W	45. 20°N66°W
4. 32°N68°W	18. 36°N86°W	32. 36°N80°W	46. 48°N72°W
5. 4°S52°E	19. 6°N66°W	33. 58°S38°W	47. 56°S15°W
6. 20°S40°E	20. 44°S77°W	34. 41°S4°E	48. 55°S54°W
7. 50°S15°E	21. 51°S5°E	35. 26°N77°W	49. 20°N72°W
8. 52°N75°E	22. 22°S45°E	36. 58° due south	50. 44°S84°W
9. 44°S15°E	23. 24°S70°W	37. 32°N85°W	51. 32°N75°W
10. 50°S49°E	24. 14°S54°E	38. 16°S48°E	52. 32°N82°W
11. 59°S23°W	25. 56°S33°W	39. 64°S74°W	53. 1°N60°W
12. 15°N68°W	26. 50°S85°W	40. 6°N62°W	54. 34°S30°E
13. 44° due W	27. 24°N73°W	41. 42°S88°W	55. 61°S33°W
14. 62°S60°W	28. 40°S28°E	42. 42°S22°E	56. 22°N76°W

57. 49°S78°W	68. 58°S88°W	79. 32°N85°W	90. 5°S55°E
58. 52°S43°W	69. 46°S27°W	80. 34°N80°W	91. 15°S50°E
59. 60°S10°W	70. 14°N63°W	81. 40°N81°W	92. 17°N72°W
60. 28°N75°W	71. 46°S25°E	82. 50°S53°W	93. 32°N73°W
61. 44°N85°W	72. 65°S24°W	83. 54°S8°E	94. 10°N64°W
62. 34°N82°W	73. 26°N78°W	84. 24°S37°E	95. 14°N71°W
63. 30°S45°E	74. 14°N66°W	85. 22°N76°E	96. 24°N78°W
64. 26°N84°W	75. 28°N72°W	86. 45°S87°W	97. 23°N68°W
65. 33°S23°E	76. 37°N83°W	87. 66°S74°W	98. 62°S45°W
66. 57°S27°W	77. 38°S88°W	88. 65°S5°W	99. 33°N78°W
67. 35°N76°W	78. 38°N78°W	89. 50°S20°W	100. 26°N70°W

233. The Upper Connoquenessing Sandstone crops out in a region with horizontal strata. A list of cross-bed dip directions is given below. Summarize these data by means of a linear histogram, a circular histogram, and an equal-area circular histogram similar to Figures 8–22, 8–23, and 8–24. What is the approximate direction of the current movement?

1. N67°W	14. S76°W	27. N59°W	39. N89°W
2. N59°W	15. N77°W	28. N74°W	40. N42°W
3. N41°W	16. S51°W	29. S87°W	41. N30°W
4. N60°W	17. N8°W	30. N49°W	42. N68°W
5. N81°W	18. N58°W	31. N5°W	43. N52°W
6. N71°W	19. N47°W	32. N63°W	44. N34°W
7. N51°W	20. N21°W	33. S70°W	45. N79°W
8. N87°W	21. S71°W	34. N3°W	46. N45°W
9. N34°W	22. N69°W	35. N26°W	47. N86°W
10. N50°W	23. N35°W	36. N66°W	48. N53°W
11. N67°W	24. N78°W	37. N72°W	49. N37°W
12. N81°W	25. N62°W	38. N46°W	50. N65°W
13. N14°W	26. N83°W		

234. At the beginning of this chapter is a list of 16 categories of geologic lineations that can be plotted on orientation diagrams. Some of these lineations are vectoral, that is, one end of the lineation can be distinguished from the other. An example is the direction of the pointed end of a straight-shelled cephalopod. Indicate which specific lineations are vectoral and which are not vectoral.

References

Bidgood, D. E. T., and Harland, W. B., 1959, Rock compass: a new aid for collecting oriented specimens: Geol. Soc. America Bull., v. 70, p. 641–644.

Billings, M. P., 1954, Structural Geology: Englewood Cliffs, N.J., Prentice-Hall, Inc., 2nd ed., 514 p.

Cox, A., and Doell, R. R., 1960, Review of Paleomagnetism: Geol. Soc. America Bull., v. 71, p. 645–768.

Curray, J. R., 1956, The analysis of two-dimensional orientation data: Jour. Geology, v. 64, p. 117–131.

Deetz, C. H., and Adams, O. S., 1934, Elements of map projection: U.S. Coast and Geodetic Survey, Special Pub. No. 68, 4th ed., 199 p.

Dennison, J. M., 1962, Graphical aids for determining reliability of sample means and an adequate sample size: Jour. Sedimentary Petrology, v. 32, p. 743–750.

Durand, D., and Greenwood, J. A., 1958, Modifications of the Rayleigh test for uniformity in analysis of two-dimensional orientation data: Jour. Geology, v. 66, p. 229–238.

Fairbairn, H. W., 1949, Structural petrology of deformed rocks: Cambridge, Mass., Addison-Wesley Publishing Co., Inc., 2nd ed., 344 p.

Flinn, D., 1958, On tests of significance of preferred orientation in three-dimensional fabric diagrams: Jour. Geology, v. 66, p. 526–539.

Freeman, J., Wise, D. U., and Bentley, R. D., 1964, Pattern of folded folds in the Appalachian Piedmont along Susquehanna River: Geol. Soc. America Bull., v. 75, p. 621–638.

Knopf, E. B., and Ingerson, E., 1938, Structural petrology: Geol. Soc. America, Memoir 6, 270 p.

Krumbein, W. C., 1939, Preferred orientation of pebbles in sedimentary deposits: Jour. Geology, v. 47, p. 673–706.

Lahee, F. H., 1961, Field geology: New York, McGraw-Hill Book Co., 6th ed., 926 p.

Nevin, C. M., 1949, Principles of structural geology: New York, John Wiley & Sons, 4th ed., 410 p.

Oertel, G., 1962, Extrapolation in geologic fabrics: Geol. Soc. America Bull., v. 73, p. 325–342.

Paterson, M. S., and Weiss, L. E., 1961, Symmetry concepts in the structural analysis of deformed rocks: Geol. Soc. America Bull., v. 72, p. 841–882.

Potter, P. E., and Pettijohn, F. J., 1963, Paleocurrents and basin analysis: New York, Academic Press, 296 p.

Raup, O. B., and Miesch, A. T., 1957, A new method for obtaining significant average directional measurements in cross-stratification studies: Jour. Sedimentary Petrology, v. 27, p. 313–321.

Schmidt, W., 1925, Gefügestatistik; Tschermaks Mineralog. Petrog. Mitt., v. 38, p. 395–399.

Spencer, E. W., 1959, Geologic evolution of the Beartooth Mountains, Montana and Wyoming. Part 2: Fracture patterns: Geol. Soc. America Bull., v. 70, p. 467–508.

Turner, F. J., and Weiss, L. E., 1963, Structural analysis of metamorphic tectonites: New York, McGraw-Hill Book Co., 545 p.

Weiss, L. E., 1959, Structural analysis of the basement system at Turoka, Kenya: Overseas Geology and Mineral Resources, v. 7, p. 133, 134.

Whitten, E. H. T., 1966, Structural geology of folded rocks: Chicago, Rand McNally and Co., 663 p.

Woolnough, W. G., and Benson, W. N., 1957, Graphical determination of the dip in deformed and cleaved sedimentary rocks: Jour. Geology, v. 65, p. 428–433.

CHAPTER 9

Faults

A STUDY OF FAULTS can be one of the most complex phases of structural geology. There are numerous descriptive terms and types of classifications applied to faults, and three-dimensional thinking is even more necessary for their analysis than in studies of most other structures. Further complications are introduced by writers who use the same or similar terms but ascribe to them different meanings. This chapter presents some of the less complex mathematical and graphical methods of fault analysis, but even these should be used selectively.

Much material related to faulting has already been considered in the preceding chapters. The determination of the apparent dip of a fault is described in Chapter 1. Plotting faults in structure sections is illustrated in Chapter 2. Methods for computing the depth to a fault are listed in Chapter 4. Chapter 5 shows how to plot the surface outcrop of a fault on a map. The illustration of faults by structure contours is presented in Chapter 6. Determination of the line of intersection of a fault plane with another planar surface is described in Chapter 8. Even Chapters 10, 11, and 12 contain concepts that can be applied to the study of faults.

The student should have a good understanding of the general principles of faulting and fault terminology before attempting a mathematical or graphical analysis of the amount and direction of movement along a fault. Useful general references on faults include the following: Nevin (1949, p. 83–145 and 372–379); Billings (1954, p. 124–241 and 455–481), Russell (1955, p. 102–155), Lahee (1961, p. 222–268), Hills (1963, p. 160–210), Badgley (1965, p. 157–277).

Fault Terminology

The concern of the present chapter is naming the components and types of motion of faults, not giving a genetic classification. Interpretation of the manner of origin of a fault should follow analysis of the type and amount of displacement. Fault displacement can be described only as relative motion of the opposing sides of the fault surface. Only rarely is enough information available to describe the actual motion of the fault blocks.

Figure 9–1 shows the relationships of several terms applied to faults. The fault surface is assumed to be a plane that includes points A, B, E, and G. Points A and E were originally coincident, and the apparent displacement of those two corners of the fault block is the true displacement. The common practice of representing the true movement of a fault by offsetting of the corners of the fault block can mislead the student. Nature does not make faults from such simple blocks; instead, the geologist strives to locate *points* on opposite sides of the fault surface that were coincident before faulting. Examples include the intersection of a distinctive stratum with a distinctive vein (creating a distinct point of intersection on both the hanging and foot wall) or the position of the axial line of a fold along a distinctive bed on both the hanging and foot wall. In Figure 9–1 points B, C, E, H, J, and K all lie in the same vertical plane. Plane AGF is vertical, as is plane ACE. Lines AB and EG are level (strike) lines on the fault. The following definitions are illustrated:

Strike of fault—the direction of a horizontal or level line on the fault plane (lines AB and EG).

Dip of fault—the direction and amount of inclination of the fault plane with respect to the horizontal, measured in a vertical plane at right angles to the strike of the fault (angle CBE in the direction BC).

Hade—the complement of the dip of the fault. Angle CEB is the hade of the fault, measured in a vertical plane normal to the strike of the fault. Hade is used chiefly as a mining term, while most geologists prefer dip as a description of the attitude of a fault.

FIG. 9–1 Illustration of fault measurements named and defined in text. Line AB is strike of fault. Fault dips in direction BC with a dip amount equal to angle CBE. Hade of fault is angle CEB. Net slip is AE; dip slip is BE. Horizontal dip slip is BC; vertical slip is CE. Angle CAE is plunge of net slip. Angle BAE is rake of net slip.

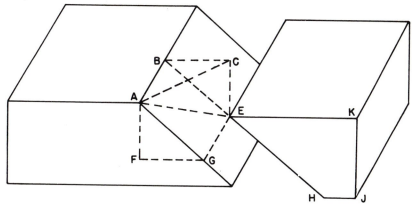

Slip—the relative displacement of formerly adjacent points on opposite sides of the fault, measured in the fault plane. Slip is a measurement of the actual relative motion of the opposing surfaces along the fault plane. (It is not *apparent* motion in a map view, in a cross section drawn normal to the strike of the fault, or in a section at right angles to the strike of the bedding cut by a fault.)

Net slip—the total displacement of the formerly adjacent points on opposite sides of the fault plane (line AE).

Strike slip—the component of net slip that is parallel to the strike of the fault (AB or EG).

Dip slip—the component of the net slip that is parallel to the dip of the fault (AG or BE).

Horizontal dip slip—the horizontal component of the dip slip or the component of the net slip that is in a horizontal position perpendicular to the strike of the fault (FG or BC).

Vertical slip—the vertical component of the dip slip or net slip (AF or CE). The relations among the types of slip values can be determined by simple geometric, trigonometric, or graphical methods. Three useful formulae are

$$\text{Net slip} = \sqrt{(\text{Strike slip})^2 + (\text{Dip slip})^2} \ . \tag{62}$$

$$\text{Net slip} = \sqrt{(\text{Horizontal dip slip})^2 + (\text{Vertical slip})^2 + (\text{Strike slip})^2} \ . \tag{63}$$

$$\text{Dip slip} = \sqrt{(\text{Horizontal dip slip})^2 + (\text{Vertical slip})^2} \ . \tag{64}$$

Plunge of net slip—The angle between the net slip and the horizontal, measured in a vertical plane (CAE is the plunge angle of the net slip; AC is the plunge direction). A formula was developed on page 112 for calculating the plunge amount:

$$\text{Tan plunge angle} = (\text{Tan true dip angle})(\text{Cos angle between true dip and plunge directions}). \tag{65}$$

Rake of net slip—the acute angle between the net slip and the strike of the fault, measured in the fault plane (angle BAE). The following formula was derived on page 112 to show the relationship between rake and plunge:

$$\text{Cos Rake} = (\text{Cos angle between plunge direction and strike of fault})(\text{Cos plunge angle}). \tag{66}$$

Normal slip fault—a fault in which the hanging wall actually did move down relative to the footwall, not merely the apparent movement of offset beds or other reference surface.

Reverse slip fault—a fault in which the hanging wall actually did move up relative to the footwall.

Strike-slip fault—a fault in which the actual movement was parallel to the strike of the fault. Other names are used for special kinds of strike-slip faults, including transcurrent fault, wrench fault, and tear fault.

Oblique-slip fault—a fault in which the net slip lies between the direction of dip and the direction of strike. Figure 9–1 is an oblique-slip fault.

Translational fault—a fault in which there has been movement of the opposing blocks along the fault surface without rotation of the blocks about an axis normal to the fault plane. All of the faults described so far have been translational.

Rotational fault—a fault in which the opposing blocks have been rotated about an axis normal to the fault plane. Some faults show a combination of translational and rotational movement. It is generally not possible to determine the relative importance of each type of motion.

Shift—the total relative displacement on opposite sides of the fault, or faults, if faulting occurs along a fault zone rather than in a single plane. In fact, sometimes the fracture is not planar, but may even be a curved surface. It is best to measure the amount of shift outside the fault zone. Shift includes the movement caused by drag-folds associated with the fault as well as the sum of the movements of the individual faults along a fault zone. Shift is calculated in the same general manner as slip. Comparable to slip, the shift can be divided into components such as net shift, dip shift, and strike shift. If Figure 9–2 is considered as a *vertical cross section* drawn normal to the strike of the fault, CH is the dip shift as compared with the dip slip (EG). Considering Figure 9–2 as a *map view*, then CH is the strike shift and GE the strike slip, if curved bedding surface ABG-EFJ is a vertical bed.

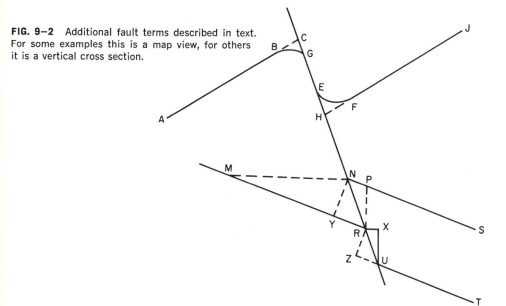

FIG. 9–2 Additional fault terms described in text. For some examples this is a map view, for others it is a vertical cross section.

Separation—the distance, measured in any desired direction, between the two parts of the disrupted horizon on opposite sides of the fault. Separation concerns apparent movement of a fault with respect to a reference horizon cut by that fault.

Horizontal separation—the separation measured in any specified direction in a horizontal plane. There are several types of horizontal separation.

Strike separation—the horizontal separation measured parallel to the strike of the fault. If Figure 9–2 is a *map view* and if MR-UT was the formerly continuous horizon, UR is the strike separation.

Dip separation—the distance between two parts of a disrupted reference plane measured in the fault plane parallel to its dip. If Figure 9–2 is a *cross section* drawn normal to the strike of the fault and if MR-UT was the formerly continuous horizon, UR is the dip separation.

Heave—the horizontal component of the dip separation, measured in a vertical *cross section* that is perpendicular to the strike of the fault. If Figure 9–2 is a vertical cross section perpendicular to the strike of the fault CU, then RX is the heave component of dip separation RU.

Throw—the vertical component of the dip separation measured in a vertical *cross section* that is perpendicular to the strike of the fault. If Figure 9–2 is a vertical cross section perpendicular to the strike of fault CU, then UX is the throw component of the dip separation RU.

Offset (normal horizontal separation)—the horizontal separation measured perpendicular to the strike of the reference horizon. If Figure 9–2 is a *map view* and if MR-UT was the formerly continuous horizon, RZ is the offset. If Figure 9–2 is considered a *cross section* normal to the strike of the bedding surface MR-NS (the formerly continuous horizon, in this example), then MN is the offset. The offset amount can be calculated in this situation by the Law of Sines:

$$MN = \frac{NR\,(Sin\ NRM)}{Sin\ NMR}. \tag{67}$$

Gap—the component of strike separation perpendicular to offset, if the offset must be determined by using a projection of the strike of the horizon. If Figure 9–2 is a *map view* and if MR-UT was the formerly continuous horizon, UZ is the gap.

Overlap—the component of strike separation perpendicular to the offset, if the offset can be determined without using a projection of the strike of the horizon. If Figure 9–2 is a *map view* and if MR-NS was the formerly continuous horizon, RY is the overlap.

Offset, gap, and overlap can be determined graphically or calculated by simple trigonometry if the amount of strike separation and the angle between the strikes of the fault and the reference horizon are known.

Vertical separation—the separation measured along a vertical line. If Figure 9–2 is a *cross section* drawn normal to the strike of the fault and if MR-NS was the formerly continuous horizon, PR is the vertical separation. Its value can be calculated from the dip separation (NR) and angles RNP and NPR by using the Law of Sines:

$$PR = \frac{NR\,(Sin\ RNP)}{Sin\ NPR}. \tag{68}$$

Perpendicular separation—the distance between the planes of two parts of a reference horizon that was a continuous plane before faulting, measured normal to the plane of the horizon. If Figure 9–2 is a *cross section* drawn normal to the strike of the formerly continuous vein MR-NS, distance NY is the perpendicular separation.

Perpendicular separation is calculated in the same manner as the thickness of a bed (Chap. 3). A point in each of the two parallel and formerly coincident planes (reference horizon) is located on opposite sides of the fault plane (that is, a point along the intersection line of the reference horizon with the hanging wall and a point along the intersection line of the reference horizon with the footwall). (These two points could be located on the ground surface where the reference horizon would intersect opposite sides of the fault on a geologic map.) The length and attitude of a line in the fault plane connecting these two points is used in the same manner as the length and attitude of the traverse line in thickness problems. The traverse slope distance is the length of the line connecting these two points in the fault plane. The traverse slope angle and direction is the plunge of this traverse line in the fault plane. The attitude of the offset horizon is used in the same way as the bedding attitude is used in thickness problems. These values can be substituted in Formulae 20 and 24.

Perpendicular separation can serve effectively to describe a fault with translational movement, but it is not a precise descriptive term for use with rotational faults, because the perpendicular separation varies along the trace of a rotational fault.

Stratigraphic throw—the thickness of the strata displaced by the fault movement, that is, the total thickness of strata that would ordinarily be present between a stratigraphic horizon at a point on one side of the fault and the stratigraphic horizon immediately adjacent to that point on the opposite side of the fault. If Figure 9–2 is a *cross section* normal to the strike of the bedding and if MR-UT was the formerly continuous stratum, RZ is the stratigraphic throw of the fault. Stratigraphic throw is a special type of perpendicular separation.

If the beds adjacent to the fault are drag-folded or if the fault movement was rotational, it is often impossible to calculate the stratigraphic throw from the information available at the fault itself. A detailed knowledge of regional stratigraphy and stratigraphic thickness data can then be used to estimate the stratigraphic thickness that ordinarily occupies the interval between the two adjacent horizons on opposite sides of the fault. At other times, the concept of stratigraphic throw is expressed by stating, for instance, that lower Ordovician is thrust upon middle Devonian.

Apparent and Actual Fault Movements

Apparent amounts and directions of relative movement along a fault can be a deceptive indicator of the actual motion (slip plus amount of rotational motion). It is for this reason that a distinction is recognized between slip and separation. The situation is exemplified by Figure 9–3, which shows a vein (AFW-OPU-EG) and a dike (BCN-WLK-SQT-JN) which are cut by the same fault. In a vertical cross section constructed in the plane RPQUT, the heave and throw based on the vein and dike differ from each other and from the true horizontal dip slip and vertical slip components of the net slip of the fault. Note also that the fault appears normal on the basis of the relative displacement of the dike (a normal separation fault), while the fault is a reverse separation fault when the vein is considered. Consideration of the true relative motion of the fault classifies it as a normal slip fault. This example illustrates that the vertical and horizontal separation are not always equal to the vertical slip and horizontal dip slip, respectively. The dike and vein have different amounts of strike separation in the plan view, both of which differ from the strike slip or actual motion of the fault block. In this example strike slip is in the same direction as strike separations of both the vein and dike.

It is worth noting again that natural situations do not consist of rectangular blocks that were slipped along a previously cut plane. Separations are apparent movements along a fault. The only way to obtain an indication of slip (true movement) in nature is to determine the orientation of the fault surface carefully and then locate a *point* on the footwall that was once coincident with another point on the hanging wall of the fault. Only the intersection point of the uniquely identifiable traces of two surfaces that intersect the hanging wall surface can form a unique point on the hanging wall, and then the corresponding unique point must be located on the footwall. It is obvious that the information necessary for solving the slip of a fault can be obtained only rarely in nature. In areas with homoclinal strata, but no dikes or veins intersecting the strata, it is generally impossible to calculate the slip of a fault without additional information such as the rake of slickensides or the assumption of a single specific type of translational fault movement (for example, assuming that all of the fault motion is dip slip motion). If the fault is rotational, it will display different dips of the reference plane (such as bedding or a vein) on opposite sides of the fault. If the possibility of a combination of rotational and translational motion exists, then no unique solution can be obtained for the fault, for reasons that will be described elsewhere in this chapter.

Intersection of a Fault and a Plane

One of the basic problems of faulting is to determine the attitude of the line of intersection of a fault with another plane, such as a vein, dike, bed, or another fault. This is simply a particular application of the general problem of finding the intersection of any two planes. An example of a fault and a vein is worked in Chapter 8 as an illustration of determining a lineation direction (see Fig. 8–2). Formulae 58 and 59 can be used to calculate the plunge amount of the line of intersection of a reference plane with a planar fault surface, and Formula 60 permits determination of the rake of that line on the fault plane.

Net Slip and Rotation along Faults

The ultimate aim of all fault description is an attempt to ascertain from the apparent displacements the true displacement along the fault, that is the net slip and amount of rotation of the opposing surfaces. This not only is necessary for speculating on the cause of faulting, but it is also practically useful in locating the position of a lost geologic phenomenon (vein, coal bed, fold axis, etc.) which should be present on the opposite side of the fault.

FIG. 9-3 Series of diagrams showing deceptive location of reference points along a fault.

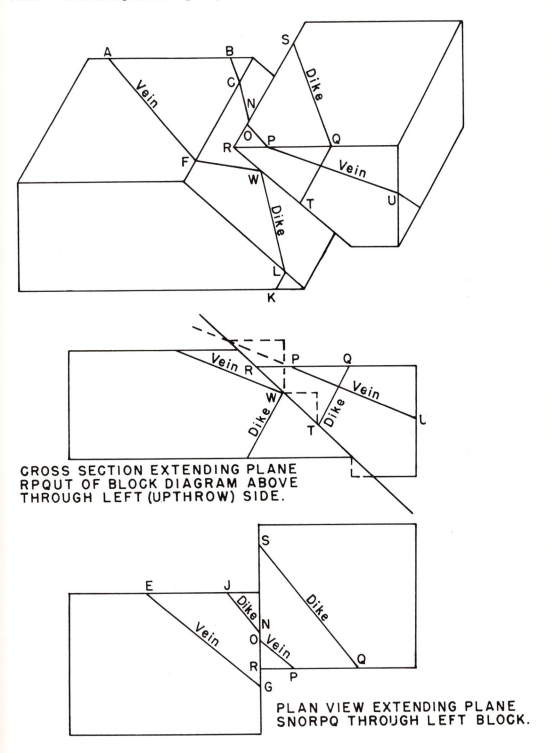

CROSS SECTION EXTENDING PLANE
RPQUT OF BLOCK DIAGRAM ABOVE
THROUGH LEFT (UPTHROW) SIDE.

PLAN VIEW EXTENDING PLANE
SNORPQ THROUGH LEFT BLOCK.

Most geologists solve fault displacement problems by means of descriptive geometry drawings in which inclined planes are rotated along fold lines until they are in plan view. Such drawings are tedious to prepare and are complex to understand, but are a long-established method of solving fault problems. This procedure is illustrated by Nevin (1949, p. 372–379), Billings (1954, p. 474–481), Donn and Shimer (1958, p. 107–123), and Badgley (1959, p. 174–186). A very detailed analysis of faults by descriptive geometry and trigonometry is presented by Haddock (1938), with special emphasis on mining problems.

The procedure outlined below involves constructing an accurate plot of the footwall geology on a sheet of paper, followed by a plot of the hanging wall geology on an overlay sheet of paper. All motion is in the plane of the fault, so the two scale drawings can be moved over one another until the structures drawn on the two sheets are congruent. The motion that produced the fault is the opposite of the motion necessary to restore the superface and subface to congruent positions. The method has the great advantage that the fault motion can be studied by trial and error, and the amounts of superface and subface motion can be readily demonstrated to an unskilled observer without explaining the subtleties of descriptive geometry.

Translational Movement / Figure 9–4 illustrates a problem of determining the net slip of a fault with translational movement only. We know that the fault lacks a rotational component because corresponding key horizons have the same attitude on both sides of the fault. Figure 9–4 is a scale-drawing plan view of a fault dipping 67° due south, a bed with an attitude of S60°W 22°NW, and a vein with an attitude of N20°W 45°NE. In addition, a dike intersects the hanging wall at point T, but the dike has not been located on the footwall. This dike has an attitude of N70°W 15°SW. It is desired first to determine the net slip of the fault, and then to locate the missing dike on the footwall.

Figure 9–5 is a block diagram showing reconstruction of the faulting motion. It is assumed that the upthrown block was eroded down to the level of the downthrown block to produce the plan view of Figure 9–4. The first step in preparing this reconstruction is to find the plunges of the lines of intersection of the fault with the bed, vein, and dike (Fig. 8–2 and Formula 58). Then convert these plunges to rake on the fault surface, using Formula 60. Information is then available to plot the traces of the bed, vein, and dike on the fault plane, as shown in Figure 9–6.

The first step in making the fault surface plots is to prepare two sheets of paper with coincident dip (parallel to the dip of the fault plane) and strike (parallel to strike of fault) reference coordinates. Then superimpose the sheets with their reference coordinates coincident, and label the top and bottom sheets hanging wall and footwall, respectively. Draw on the sheets, with coordinates matched, a horizontal line (ABCET of Fig. 9–6) corresponding to the strike of the surface outcrop of the fault. On the footwall sheet plot the points of intersection of the vein (A) and bed (C). Then plot the traces of the vein (AL) and bed (CF) by using the rake angle for each. Lines on the footwall are shown as solid lines in Figure 9–6.

The same procedure should be used for the hanging-wall sheet. The traces of the

FIG. 9–4 Map illustrating method of delineating net slip along a fault surface.

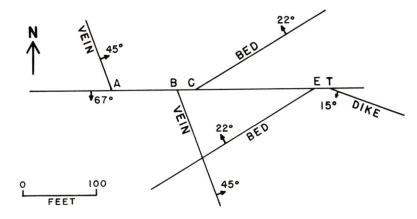

FIG. 9–5 Diagram of fault block which was eroded down to the level of line AYBCET to produce the situation shown in the map of Figure 9–4.

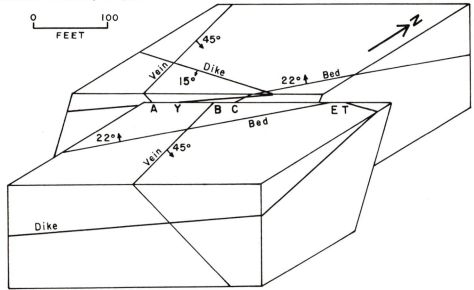

FIG. 9–6 Projection into the fault plane of the situation mapped in Figure 9–4.

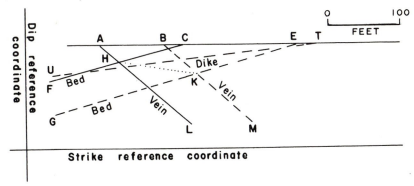

vein (BM), bed (EG), and dike (TU) are plotted as dashed lines on the overlay or hanging-wall sheet.

The movement necessary to make the hanging wall and footwall traces of the bed and vein coincide is a restoration of the slip that produced the translational fault. After placing the traces of the vein and bed coincident with each other (Fig. 9–7), the

FIG. 9–7   Translation of Figure 9–6 to remove effect of faulting.

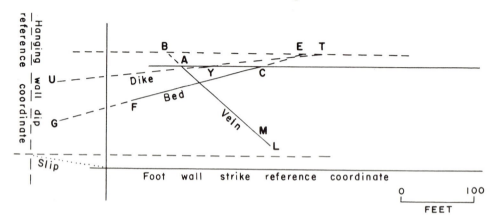

dotted line drawn connecting the points of intersection of the strike and dip reference coordinates on the hanging-wall and footwall sheets is the relative direction of motion and amount of net slip in the fault plane. Of course, the fault motion necessary to restore the displaced parts is in exactly the opposite direction from the fault motion that produced the map-view situation of Figure 9–4. Net slip also is given by the offset point of intersection of the bed and the vein on the hanging wall and footwall. Both of these intersections are visible in Figure 9–6, where the net slip amount is dotted line HK and the rake is the angle between HK and AE. The rake of the net slip (determined either from Fig. 9–6 or 9–7) can be converted into plunge of the net slip by the following equation:

$$\text{Sin plunge angle} = (\text{Sin rake})(\text{Sin dip angle of fault}). \tag{69}$$

$$\text{Sin} = \left(\begin{array}{c}\text{angle between strike of fault and}\\ \text{plunge of net slip directions}\end{array}\right) = \frac{\text{Tan plunge angle of net slip}}{\text{Tan true dip angle of fault}}. \tag{70}$$

The length of the net slip plus its plunge direction and amount can be used to calculate strike slip, dip slip, horizontal dip slip, vertical slip, and other expressions of relative movement, using simple trigonometry or scale drawings.

The problem still remains to locate on the footwall the trace of the dike that forms line TU on the hanging wall of Figure 9–6. The same line is shown in pre-faulting position on the hanging-wall portion of Figure 9–7. Place the footwall sheet above the hanging-wall sheet exactly as in Figure 9–7, and then trace onto the footwall the pre-faulting position of the dike, extending the dike somewhat beyond the limits plotted on the hanging-wall sheet. The dike intersects line AC on the footwall at point Y. The distance AY can be transferred to Figure 9–4, locating on the plan view the intersection of the dike with the footwall of the fault. The strike of the dike on the plan view of the footwall would be the same as the strike of the dike on the hanging wall, because the fault is nonrotational.

Rotational Fault Movement / Rotational faults involve a rotation about some axis normal to the fault surface (known as the *normal pole*). A rotational fault is readily distinguished because the attitude of the planar reference horizons is different in the footwall and hanging-wall blocks (Fig. 9–8). Once the normal pole is located, it is easy to find the angle through which the fault has been rotated.

The displacement along any rotational fault can be explained simply by rotation about a uniquely located normal pole, provided that the geologist is willing to assume that the displacement is caused totally by rotation and that there has been no translation movement. The plan view of Figure 9–8 is the result of the movement indicated in Figure 9–9, according to information developed in a diagram of the fault plane (Fig. 9–10).

In Figure 9–10 the hanging-wall and footwall traces of the bed intersect at point A, and point E is the corresponding intersection of the dike traces. Angles BAC and FEG are both 30°, which is the amount of rotation about the normal pole that is necessary for restoring the continuity of the bedding and dike traces on the hanging wall and footwall. Point T is the position of the normal pole about which the rotation takes place. To find that normal pole, draw line AE, then bisect it with the perpendicular line at R. Point T, the axis of rotation, is located by trial and error along line RT, using a pin as an axis for rotating the hanging-wall sheet over the footwall sheet. In this example, the axis of rotation happens to intersect the top of the hanging-wall block at the fault trace (see Fig. 9–9).

The angle of rotation can be checked independently by the stereonet method described on pages 182–183. The stereonet solution tells only the amount of rotation about the normal pole, not the location of the point where the normal pole intersects the fault plane.

Translational plus Rotational Movement / If the possibility exists that a fault has experienced translational movement in addition to rotational movement, then a unique solution for the total motion is not possible. The angle of rotation will be the same whether or not translational movement occurred, but the location on the fault surface of the normal pole of rotation will shift if translational movement has occurred.

Figure 9–11 shows the plot of the fault plane in Figure 9–10, except that the hanging wall has been moved down-dip 200 feet and shifted to the right (northeast) along strike 100 feet relative to the footwall. Point T no longer is the axis of 30° rotational movement. Instead, trial-and-error determination reveals that point S is the pivot about which rotation of the hanging-wall sheet must occur to allow lines BH and KF to become superimposed on AC and EG, respectively.

The difficulty in rotational fault interpretation results from the fact that a unique

FIG. 9–8 Plan view of rotational fault.

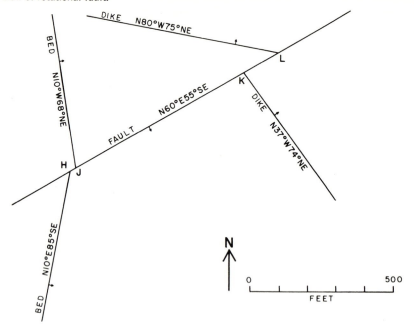

FIG. 9–9 Fault movement that produced plan view in Figure 9–8, after eroding footwall block down to level of top surface of hanging-wall block.

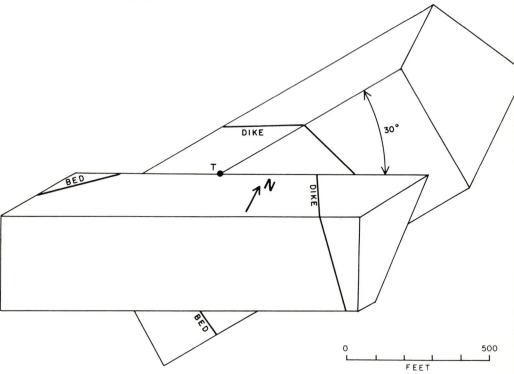

FIG. 9–10 View in fault plane, showing movement that resulted in plan view of fault as shown in Figure 9–8. Movement has been purely rotational about point T, with no translation motion.

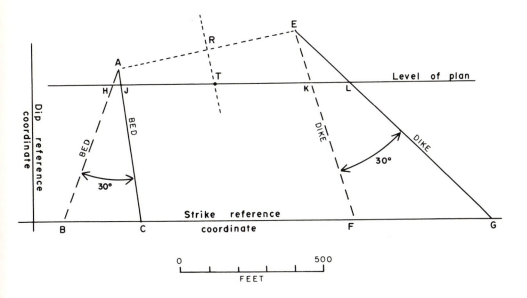

FIG. 9–11 Combination of rotary and translatory motion which could produce the plan view of Figure 9–8 after erosion. Hanging wall rotated 30° counterclockwise about point S and then moved 200 feet up dip and 100 feet to left (southwest).

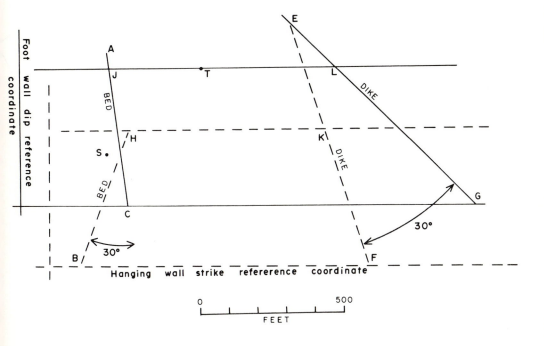

normal pole can be located for zero translation or for translation with any specified amount of strike-slip and dip-slip motion. Unfortunately it is not possible to detect from a map of the fault with plotted dips of the reference planes, whether any or how much translational motion occurred in addition to rotary motion. If additional assumptions are made, then a unique solution of pole position and angle of rotation is possible. In Figure 9–11, by assuming 200 feet of dip-slip and 100 feet of strike-slip motion in a specified direction, then the unique normal pole of rotation can be specified by trial and error search. Similarly, if a person is willing to assume that all rotation is about point K, then the angle of rotation can be calculated (30°) and the strike slip and dip slip can be uniquely determined. (The sequence of movements could be rotation counterclockwise of 30° about point K on the hanging wall, followed by 160 feet of up-dip motion and 40 feet of strike-slip motion to the northeast; of course, all of these movements could take place simultaneously.)

Irregular Topography / The solutions illustrated in Figures 9–8 through 9–11 assume that the map of Figure 9–8 is a horizontal plane surface and that the fault trace is a horizontal line. In areas with irregular topography, ground elevations along the fault trace must be plotted as dip-direction distances perpendicular to the strike reference coordinate, measured in the plane of the fault. Convert the vertical distance above the strike reference coordinate to dip-direction distance by multiplying the elevation difference times the cosecant of the true dip angle of the fault. Traces in the fault plane of reference horizons are plotted by using the outcrop position of the reference plane (such as a dike intersecting the fault in map view) and the rake angle in the fault surface.

Curved Faults / The procedures described in this chapter are designed for planar fault surfaces. They become increasingly inaccurate as the fault surface deviates from a plane. The method presented also assume that all other references (such as dikes, beds, or veins) are planar surfaces. If the traces of non-planar reference surfaces intersecting the fault plane are known accurately, the method of hanging-wall overlay sheet still may be used. The problem becomes quite complex when the effect of drag is also considered. It is possible to obtain approximations of the relative fault movement, however, if the relations are plotted with care.

In actual practice, the chief difficulty with faults is that the information necessary for solving problems is difficult or often impossible to obtain. The geologist should be careful that he has sufficient data for solving a specific problem.

Exercises /

235. Use the map of Figure 9–4 as the base for this exercise, except that the dip amounts are changed for the fault, vein, and bed. Label the bed as dipping 37°N30°W, the vein dipping 65°N70°E, and the fault dipping 40° due south. The dip of the dike is as indicated in Figure 9–4. Determine the amount of net slip, strike slip, amount and direction of plunge of net slip, rake of net slip,

horizontal dip slip, and vertical slip of the fault. Calculate the strike separation, offset, gap, and perpendicular separation of the vein. Compute the strike separation, offset, overlap, and stratigraphic throw of the beds associated with the fault. Is the fault normal slip or reverse slip? Locate the intersection trace of the dike on the footwall, and plot the outcrop of the dike at its expected position on the footwall block. Assume that Figure 9–4 is a plan view.

236. Figure 9–12 is a map view with a scale of 1:12,000. Determine the rotational movement and plunge and amount of net shift of the fault with respect to point B on the footwall. Determine the strike shift and vertical and horizontal components of the dip shift with respect to point B. Determine the strike separation of the vein and the bed. Is the fault normal or reverse?

237. This exercise is the same as 236 except that point A on the footwall should be assumed as the normal pole for rotation, and then have translation movement to produce the present outcrop pattern on the hanging wall. Characterize the amount of rotation and the components of the net shift motion. Is the dip-slip component of translation motion normal or reverse?

238. Figure 9–12 is a map view with a scale of 1:12,000. Determine the location of the point (on the fault surface projected into space) about which the total motion of the fault appears rotational. Locate this normal pole in the plane of the fault with respect to point B, using the strike of the fault passing through B as an abscissa and a dip line passing through B as an ordinate. What rotation about this point could produce the configuration shown in Figure 9–12, assuming that the footwall did not move?

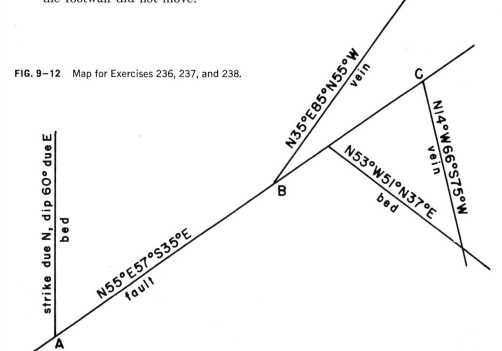

FIG. 9–12   Map for Exercises 236, 237, and 238.

239. Figure 9–13 is a plan view of a fault. Determine the location of the normal pole for the rotational movement, assuming that there has been no translation movement. How many degrees has the hanging wall rotated relative to the footwall? Locate the expected trace of the vein on the footwall and the point on the map where the vein should intersect the fault trace on the footwall.

FIG. 9–13 Map for Exercise 239.

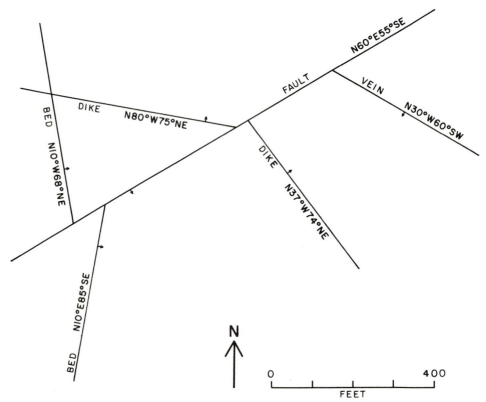

240. Figure 9–14 is a plan view of a translational fault. The fault plane is vertical. Determine the strike slip and dip slip. Thicknesses of the various formations are as follows: A, 30 feet; B, 50 feet; C, 20 feet; D, 40 feet; E, 30 feet; F, 50 feet; G, 70 feet. Dips are as indicated on the map. Is the fault a right lateral or a left lateral strike slip fault?

FIG. 9–14   Plan view of fault for use with Exercise 240.

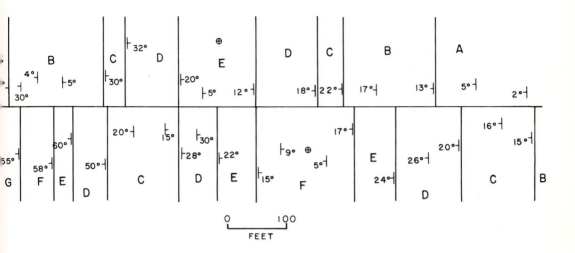

References

Anderson, E. M., 1951, The dynamics of faulting: London, Oliver and Boyd, 2nd ed., 206 p.

Badgley, P. C., 1959, Structural methods for the exploration geologist: New York, Harper and Bros., 280 p.

———, 1965, Structural and tectonic principles: New York, Harper and Row, 521 p.

Beckwith, R. W., 1947, Fault problems in fault planes: Geol. Soc. America Bull., v. 58, p. 79–108.

Billings, M. P., 1954, Structural geology: Englewood Cliffs, N.J., Prentice-Hall, Inc., 2nd ed., 514 p.

Crowell, J. C., 1959, Problems of fault nomenclature: Am. Assoc. Petroleum Geologists Bull., v. 43, p. 2653–2674.

Dickenson, G., 1954, Subsurface interpretation of intersecting faults and their effects on stratigraphic horizons: Am. Assoc. Petroleum Geologists Bull., v. 38, p. 854–877.

Donath, F. A., 1964, Folds and faulting: Geol. Soc. America Bull., v. 75, p. 45–62.

Donn, W. L., and Shimer, J. A., 1958, Graphic methods in structural geology: New York, Appleton-Century-Crofts, Inc., 180 p.

Gill, J. E., 1941, Fault nomenclature: Royal Soc. Canada Trans., 3rd series, sect. IV., v. 35, p. 71–85.

Haddock, M. H., 1938, Disrupted strata: London, The Technical Press, Ltd., 2nd ed., 104 p.

Hill, M. L., 1959, Dual classification of faults: Am. Assoc. Petroleum Geologists Bull., v. 43, p. 217–221.

Hills, E. S., 1963, Elements of structural geology: New York, John Wiley & Sons, 483 p.

Kupfer, D. H., 1960, Problems of fault nomenclature: Am. Assoc. Petroleum Geologists Bull., v. 44, p. 501–505.

Lahee, F. H., 1961, Field geology: New York, McGraw-Hill Book Co., 6th ed., 926 p.

Moore, C. A., 1963, Handbook of subsurface geology: New York, Harper and Row, 235 p. (Chapter 8 considers criteria for recognizing faulting in petroleum wells.)

Nevin, C. M., 1949, Principles of structural geology: New York, John Wiley & Sons, 4th ed., 410 p.

Russell, W. L., 1955, Structural geology for petroleum geologists: New York, McGraw-Hill Book Co., 427 p.

Threet, R. L., 1964, Trigonometric scales to simplify descriptive geometry in structural geology: Jour. Geological Education, v. 12, p. 126–129.

CHAPTER 10

Dip Determination from Drill Cores

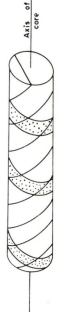

THE CYLINDER OF ROCK RECOVERED AS A DRILL CORE is often intersected by planar structures such as bedding, schistocity, veinlets, or joints (Fig. 10–1). Unfortunately the original attitude of these planar structures within the hole is not known, because the pieces of core have rotated about the axis of the hole during drilling and removal operations. It is possible to obtain a surprising amount of information about the strike and dip of these planar structures using core data, even though a complete solution for strike and dip usually is not possible without three cores intersecting the structure.

Types of Data

The planar structure intersecting the core can be a specific plane, such as the top of a key bed, or it can be a series of parallel planes, none of which is distinctive enough to be a key horizon. Schistocity, cleavage, a joint set, or a series of similar beds would constitute this latter situation. Figure 10–1 shows sandstone beds, none of which is distinctive, that are intersected by closely spaced joints that intersect both the sandstone and intervening shale layers.

FIG. 10–1 Drill core showing several beds of sandstone and shale intersected by closely spaced joints.

Specific Plane / If a specific plane can be identified in the core, then that plane can be located at a definite point in space by using the position of the *collar* (top of the drill hole), the direction of drilling, and the distance drilled to intersect the plane. Identification of this distinctive planar surface (an unusual bed, a particular vein, etc.) in three cores permits solution of a three-point problem for figuring the strike and dip of the plane (Chap. 5). Figure 10–2 shows the procedure for using cores to determine the position of three points on a distinctive surface. Plot the map location of the collar of each drill hole. Draw a line in the direction of plunge of each inclined hole, making the length of the line equal to:

(Distance drilled) (Cos plunge amount of hole).

The lower end of this line is the map position of the intersection of the core and the distinctive plane. The elevation of the plane where intersected by a particular core equals:

(Collar elevation) — (distance drilled) (Sin plunge amount).

Three points in space determine the position and attitude of a plane, and its strike and dip can be ascertained according to methods described on pages 2–4 and 62. If hole 2 in Figure 10–2 had been vertical, the intersection of the hole and plane would have been directly beneath the collar and point 2' would have coincided with point 2.

It is also possible to determine the position and attitude of a distinctive bed identified in only one core, if the angle between bedding and the axis is known for three differently inclined cores. The attitude of the key bed can be determined from the three cores by using procedures described later in this chapter, and a plane with that attitude can be constructed to pass through the position of the key bed in the one core.

FIG. 10–2 Use of drill cores to determine the data needed to solve a three-point problem. The map position of the collar is at the back of the arrow, and each tip corresponds to the map position of the intersection of a specific horizon with a core.

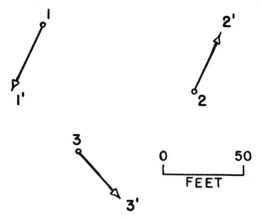

Series of Parallel Planes / A common problem is determining the attitude of a series of parallel planes, none of which is distinctive. Three differently inclined cores will yield a unique solution, assuming that the dip is the same at the three localities. In certain situations two cores are sufficient to determine a unique attitude for the planar orientation.

A plane may intersect the core axis at any angle. If the core is rotated about its axis, the apparent inclination of the planar structure changes in a manner analogous to true and apparent dip. The maximum value of the apparent angle between the axis and the plane is the true angle between the line and plane. The planar surface appears as a straight line crossing the core when the core is oriented for measuring this angle (Fig. 10–3). This angle forms a cone when rotated about the axis (Fig. 10–4). For a single core marked by a series of parallel planar structures, the total information known is that the planar structure must be oriented in the earth so that some line in the plane is an element of this cone.

FIG. 10–3 True measurement of angle between core axis and bedding.

FIG. 10–4 Cone generated by rotating planar surface about core axis.

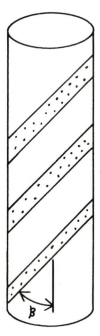

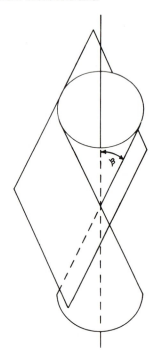

Vertical Cores

The dip amount of a planar structure encountered by a vertical core obviously equals the complement of the angle between the plane and the core axis. The strike direction is unknown when a single core is available, but the direction can be determined from two cores. The following problems illustrate the procedure.

Two vertical cores encounter a distinctive hematite bed. Core B is 175 feet due east of core A, with the two collar elevations equal. The cores are drilled on level ground. The angle between bedding and the core axis (hade) is 63° in both cores. Core A encounters the hematite bed at a depth of 24 feet, and core B penetrates it at 50 feet. What is the attitude of the bed?

The true dip amount is obviously 27°, the complement of the hade. Prepare scale drawings in plan and cross section (Fig. 10–5). In cross section lay off distance AB, and draw vertical holes with scaled depth AH and BJ. Find point E by laying off angle BJE equal to 63°. Complete the other side of the cross section of the cone by drawing FJ in a similar manner. Then draw angles CHA and DHA for the smaller cone. On the plan view prepare a map of points A and B and circumscribe them with a circle with a radius equal to AC and BE, respectively. The strike of the bed potentially could be anywhere along the circumference of each circle. However, the two tangent lines are the only possible strike directions that are common to both circle A and B. It is impossible to determine more specifically which is the true strike of the bed across the map, so there is a 50% risk of missing the bed by digging along either possible strike line. Fortunately, the possible strike lines intersect at point T, so an excavation there will be certain to expose the bed, and its true attitude can be measured directly.

If the regional strike is known fairly confidently, then two vertical holes located along a line at right angles to strike will yield a unique solution for dip direction and amount. Let us modify the preceding problem slightly.

Regional strike is due north. Core B is drilled 175 feet due east of core A, with the elevations of the collars equal. The angle between the bedding and the vertical core axis is 63° in each core. Core A encounters the hematite bed at a depth of 24 feet, and core B penetrates it at 115 feet. What is the attitude of the bed?

FIG. 10–5   General situation with two vertical drill cores intersecting distinctive bed.

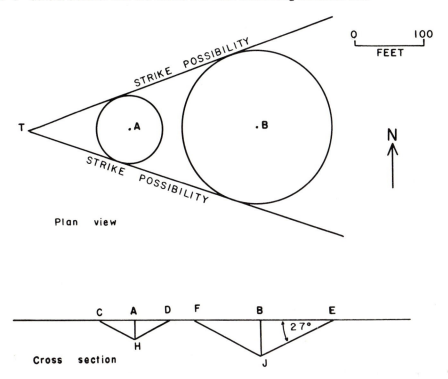

Prepare Figure 10–6 showing a scale drawing of this situation, following the steps for constructing Figure 10–5. Note that there is only one possible tangent common to both circles, so the strike direction is perpendicular to a line connecting A and B. (Recall that the holes in the example were deliberately located perpendicular to regional strike; of course, it is possible that in an ordinary drilling program two holes would chance to be located perpendicular to an unknown strike direction.) In the cross section angle BCJ is the only possible dip direction, and the dip is 27° due east. The bed can be expected to crop out at point C and along the strike line. Note that the situation for Figure 10–6 is simply a special case of the conditions in Figure 10–5.

FIG. 10–6 Special situation with two vertical drill cores whose collars are along a line perpendicular to strike.

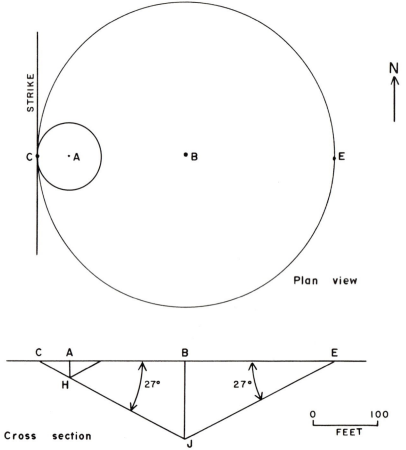

If two drill holes should happen to encounter the distinctive bed at the same elevation, then the two circles in Figure 10–5 would have an identical radius. The strike determined by the two tangent lines would be parallel to a line connecting the collar of the two holes in plan view, and it would not be possible to ascertain the true dip direction more specifically than one of the two directions perpendicular to strike. Furthermore, the tangent lines would be parallel, so there is no point where the two possible strike lines will intersect. Fortunately this situation is rare and would only be encountered by chance. If the geologist has any idea of a regional strike direction, the two holes should be drilled at locations other than with collars along a line parallel to strike, so as to avoid this possible complication.

If a marker bed cannot be identified in the cores, it is still possible to determine the attitude of the bedding planes, but not the outcrop trace of any particular bed. Plot the two holes some unit depth (say, arbitrarily, 100 feet), and draw the cross section of two cones in similar fashion, making the angle of taper of the cones correspond to the angle between bedding and the core axis. The distance between core axes is plotted to scale. Draw the same situation in plan view. Lines tangent to the circles define the strike direction, but a choice cannot be made between the two possible dip directions, unless a third and inclined hole is drilled.

Inclined Cores

If two cores inclined differently are available, it is possible to draw the cones with a common vertex in plan view in order to study the attitude of the bedding. If the attitude of a particular plane can be determined, its position is known automatically from its location in the drill hole, so that a complete description of the unique plane in space is possible. Figure 10–7 illustrates the general situation for determining the attitude of a plane using two inclined drill holes. The planar structure must lie tangent to both cones. A maximum of four possible tangent planes can be drawn, each formed in Figure 10–7 by point A and one of the four horizontal lines tangent to the top of the cones. The four horizontal lines (BB', CC', EE', FF') are possible strike lines in each of the four possible tangent planes. The line of tangency of a plane with each cone determines an apparent dip line of that plane along an element of the cone. These two apparent dips can be solved for true dip using the methods of Chapter 1 or by a simpler method described later in this chapter. A maximum of four possible dip values can be obtained. In the general case of Figure 10–7 the inclination of the planar structure cannot be determined more specifically than one of these four possibilities. The attitude of the planar structure can be determined more specifically under certain conditions.

FIG. 10–7 General case for two cores inclined differently.

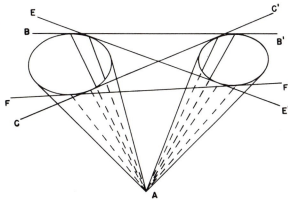

Figure 10–8 shows some of the other possible loci determined by the intersection of a horizontal plane with two inclined cones. In that figure most of the sets of two cones are inclined in the same direction, but are inclined different amounts. This was done to conserve space; actually the cones may be inclined in any direction. The two cones have a common apex whose position in plan view is indicated by a black dot. There can be three (Fig. 10–8, part a), two (Fig. 10–8, part b), or one (Fig. 10–8, part c) possible strike directions, in addition to the more general case of four. If one of the holes is vertical, its cone forms a circle rather than the more general ellipse. The special case in which one core axis is vertical has been considered by Mead (1921), Lobeck (1958, p. 131–138), and by Donn and Shimer (1958, p. 88–93). Parabolas (Fig. 10–8, parts d, e, f, g, k, l) and hyperbolas (Fig. 10–8, parts h, i, j, k, l) may also occur as conic sections. These situations have been considered by Mead (1921). Conditions under which the different conic sections occur can be summarized as follows (see Fig. 10–9):

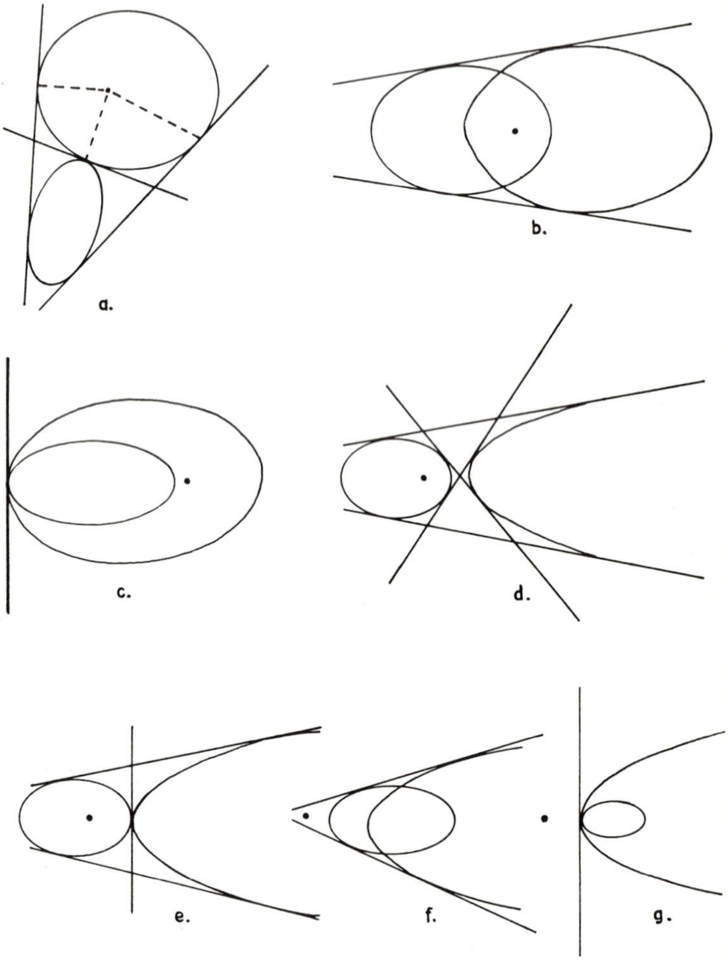

FIG. 10–8 Special cases for two cores inclined differently, shown in plan view.

Ellipse—angle between planar structure and core axis is less than plunge amount of axis. If the core axis is parallel to the planar structure, the ellipse is reduced to point size, because the apical angle is zero. The ellipse becomes a circle if the core is vertical.

Parabola—angle between planar structure and core axis equals plunge amount of axis. This special case is intermediate between an ellipse and a hyperbola.

Hyperbola—angle between planar structure and core axis exceeds plunge amount of axis. Because the cores are usually inclined more steeply than 45°, this situation occurs usually with gently dipping planar structures. A planar structure perpendicular to a drill hole is a special case of a hyperbola. ·

The following procedure is used to solve geometrically for possible attitudes of the planar structure, using two differently inclined drill holes. Place a point on the paper to represent a map view of the apex common to the two cones. This apex is actually located a unit distance below the horizontal reference plane of the paper, but viewed in plan it is plotted on the paper. For each core draw a line from the apex point in a

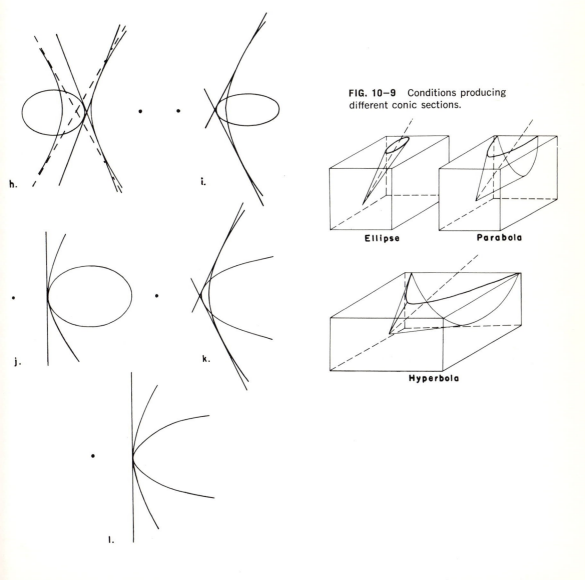

FIG. 10–9 Conditions producing different conic sections.

Ellipse

Parabola

Hyperbola

direction opposite to the plunge of the core and with

length of line = cot plunge amount of core axis.

The end of this line is the point at which the core axis pierces the horizontal reference plane (represented by the sheet of paper). The direction of the line is the major axis of the conic section. Draw to scale the proper conic section for each core. The procedure for a circular section (core axis vertical) was described at the beginning of this chapter. Construction of an elliptical section is described in detail subsequently, and references will be given for drawing a parabolic and hyperbolic section. After completing the scale drawings of conic sections, draw lines in plan view tangent to the conic sections, as in Figure 10–8. Each tangent line (or plane formed by the line plus the apex of the cone) represents one possible attitude of the planar structure. Possible true dip directions are determined by drawing lines from the point representing the common apex of the cones perpendicular to the possible strike lines (like the dashed lines in Fig. 10–8, part a). The length of such a dashed line is the cotangent of the dip amount. Follow this procedure to determine the amount and dip for all possible values of true dip.

A solution can always be obtained for two cores unless one cone lies entirely within the other and the cones touch only at the vertex. Such a situation means either that a mistake has been made in the solution or that the dip of the structure is not the same in the two cores. If the dip does differ in the two cores (the dip is not homoclinal), an apparently valid solution can usually be obtained, and it will be impossible to detect that the dips are different.

Constructing an Ellipse / It now remains to show details of the construction of conic sections resulting from the intersection of any cone with a horizontal plane. In most drill core situations the conic section generated will be a circle (axis vertical) or an ellipse (angle between planar structure and core axis is less than plunge amount of axis).

Figure 10–10 shows the construction of an ellipse. The example is for a cone plunging 40° due east, with bedding making an angle of 18° with the core axis. The plunge direction of the core axis is the direction of the major axis of the ellipse. Draw a cross section of the cone viewed from a position perpendicular to the major axis (Fig. 10–10).

1. Draw AE, the trace of the horizontal plane.
2. Draw EG perpendicular to AE. Make EG some arbitrary unit length in the scale drawing. If ellipses result from more than one core, all elliptical constructions should be drawn to the same scale by keeping EG the same length in all drawings. Draw the inclined core (FG) and the ± angle the beds make with the core axis (angles FGA and FGC). Points A and C lie on the ellipse at the major axis.
3. Bisect AC at point H. Point H is the center of the ellipse.
4. Draw IG perpendicular to FG through G.
5. Draw ML perpendicular to FG through H.
6. Draw HK parallel to FG through H.
7. Draw arc PQ about point G with radius GQ equal to ML.

8. Distance QK is the length of the semi-minor axis.
9. Distance AH is the length of the semi-major axis.
10. From the view of Figure 10–10 obtain the data to transfer the major axis of the ellipse to the map view (Fig. 10–11). The major axis lies in the direction of plunge of the core, and point E is directly above the apex of the cone in this example. Plot the major and minor axes.
11. Draw the outline of the ellipse by means of the trammel method as follows:
   On a straight edge of paper mark off the distance OS equal to the semi-minor axis, and on the same side of O mark off the distance OT equal to the semi-major axis (Fig. 10–11). Then lay this trammel across the ellipse axis so that point S is on the major axis and point T is on the minor axis. Point O lies on the ellipse. By shifting the trammel while keeping T on the minor axis and S on the major axis, one can plot a locus of points on the ellipse defined by the motion of point O on Figure 10–11.

FIG. 10–10 First steps in constructing an ellipse generated from an inclined drill core situation.

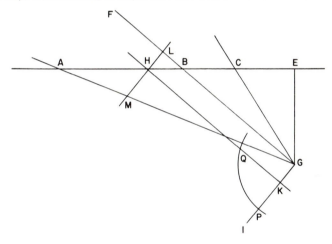

FIG. 10–11 Construction of an ellipse by the trammel method.

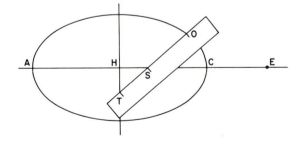

Other Conic Sections / A hyperbola or parabola is uncommon as a conic section resulting from drilling into the ground, but may frequently occur in inclined holes drilled into a mine face. The construction of these figures is not explained in detail because the procedure is complex and a much simpler solution can be obtained using stereonets, as described in the next chapter. Many descriptive geometry or engineering drawing texts describe the construction of these figures (Rule and Coons, 1961, p. 139–141; Hoelscher and Springer, 1956, p. 4–11 through 4–13).

Solution with Three Cores / Three cores with different inclinations provide a unique solution for strike and dip (Fig. 10–12). Points A, B, and C are the intersection of the horizontal plane with the three inclined core axes, all three of which intersect at a common point G. Line EHF is the only line that can be drawn tangent to all three cones. The true dip amount of the plane whose strike is EHF can be obtained by drawing a line from G perpendicular to EHF (dashed line). The length of this perpendicular line (GJ) is the cotangent of the dip amount if Figure 10–12 is drawn to scale with the apex of the three cones at a point located a unit distance beneath the plane of the horizontal surface.

 If the data from three differently inclined cores do not permit a single line to be drawn tangent to all three conic sections, this can result from an error in reading the angle of the plane with respect to the core axis, a mistake in plotting the conic sections, or else the dip of the planar structure is not the same in all three cores. It is unlikely that the data can be read and plotted accurately enough to give a line exactly tangent to all three conic sections, so often a "best fit" solution must be used. This is still reliable

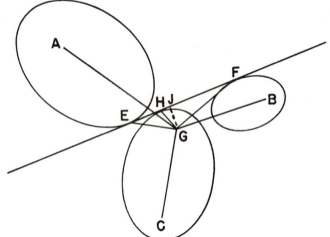

FIG. 10–12 Map-view plotting of three cores having different inclinations.

enough for most purposes. A large change of dip can always be detected with three differently inclined cores.

 The three cores need not have totally different inclinations if the special situation of two vertical cores is obtained. Recall that two vertical cores always yields a true strike direction. A third core inclined from the vertical yields sufficient information for a unique solution. The solution will be more precise if the third core plunges nearly perpendicular to a line connecting the collars of the vertical holes. Wisser (1932) presented a graphical solution for this situation.

Approaches besides Scale Drawings

The method described in the preceding pages of making scale drawings of conic sections generated by drill core data provides a readily visualized solution to the problem of determining strike and dip, but it is very tedious. Two other approaches have been described, namely trigonometric solutions and stereonet solutions.

Trigonometric formulae for solving two-core problems were developed by Haddock (1938, p. 16–22) and Stein (1941). Mertie (1943) found a trigonometric solution for three inclined cores. Lyons (1964) used a computer to prepare graphs of trigonometric solutions so that the attitude of a distinct plane (such as a marker bed) can be uniquely determined graphically, without making trigonometric calculations or scale drawings.

Stereonet solutions to drill-core problems are described on pages 184–186. Stereonet procedures were developed by Fisher (1941), Bucher (1943), and Gilluly (1944).

Mertie (1943) considered calculating the configuration of a non-planar horizon using core data.

Exercises / The following problems involve construction of elliptical or circular conic sections on a level terrain, except Exercise 242.

241. Two vertical cores start at the same elevation. The first intersects a key bed at a depth of 182 feet. The second core was taken 320 feet S56°E from the first, and it encountered the key bed at a depth of 97 feet. Bedding makes an angle of 43° with the axis of each core. What are the possible dips of the bed? What distance and in what direction should one go from the second core in order to find the single point that is the outcrop of the key bed at the same elevation as the start of the cores, and which marks the intersection of all possible strike directions?

242. The top of a certain formation makes an angle of 49° with the axis of each of two vertical cores. Core one starts at elevation 1,356 feet and encounters the contact at 300 feet depth. Core two is located 170 feet S17°E of core one; it starts at elevation 1,346 feet and intersects the contact at 140 feet depth. What are the possibilities for the dip of the bed?

243. Core one is taken with a plunge of 52°S29°W. Another core is vertical. The bedding makes an angle of 18° with each core axis. What is the attitude of the bedding?

244. Two holes are drilled into a level ground surface. Hole A plunges 64°N38°W and cleavage makes an angle of 32° with the core axis. Hole B is vertical, and cleavage makes a 64° angle with the core axis. Determine the strike and dip of the cleavage.

245. Two cores intersect slaty cleavage. Core A plunges 68°N56°E and the cleavage makes an angle of 37° with the core axis. Core B plunges 55°S23°E and the cleavage parallels the core axis. What are the two possible dips of the cleavage?

246. Core A plunges 52°S40°E, and the bedding makes an angle of 18° with the axis. Core B plunges 68°N89°W, and the bedding makes an angle of 37° with the core axis. What are the possible dips of the bed?

247. Core A plunges 52°S54°W, and the bedding makes an angle 18° with the core

axis. Core B plunges 61°S80°E, and the bedding makes an angle of 34° with the core axis. What are the possible dips of the beds?

248. Three cores intersect a series of bedding planes with the following relations:

CORE ONE: axis plunges 82°S37°W; bedding makes angle of 17° with axis.

CORE TWO: axis plunges 61°S21°E; bedding makes an angle of 34° with axis.

CORE THREE: axis plunges 50°S7°E; bedding makes an angle of 30° with axis.

What is the attitude of the bedding?

References

Bucher, W. H., 1943, Dip and strike from three not parallel drill cores lacking key beds (stereographic method): Econ. Geology, v. 38, p. 648–657.

Donn, W. L., and Shimer, J. A., 1958, Graphic methods in structural geology: New York, Appleton-Century-Crofts, Inc., 180 p.

Earle, K. W., 1934, Dip and strike problems: London, Thomas Murby and Co., 126 p.

Fisher, D. J., 1941, Drill hole problems in the stereographic projection: Econ. Geology, v. 36, p. 551–560.

Gilluly, J., 1944, Dip and strike from three not parallel drill cores lacking key beds: Econ. Geology, v. 39, p. 359–363.

Haddock, M. H., 1938, Disrupted strata: London, The Technical Press, Ltd., 2nd ed., 104 p.

Hoelscher, R. P., and Springer, C. H., 1956, Engineering drawing and geometry: New York, John Wiley & Sons, 498 p.

Johnson, C. H., 1939, New Mathematical and stereographic net solutions to problem of two tilts—with applications to core orientation: Am. Assoc. Petroleum Geologists Bull., v. 23, p. 663–685.

Lobeck, A. K., 1958, Block diagrams and other graphic methods used in geology and geography: Amherst, Mass., Emerson-Trussell Book Co., 2nd ed., 212 p.

Lyons, M. S., 1964, Interpretation of planar structure in drill-hole core: Geol. Soc. America, Special Paper 78, 65 p.

McClellan, H., 1948, Core orientation by graphical and mathematical methods: Am. Assoc. Petroleum Geologists Bull., v. 32, p. 262–282.

Mead, W. J., 1921, Determination of attitude of concealed bedded formations by diamond drilling: Econ. Geology, v. 16, p. 37–47.

Mertie, J. B., Jr., 1943, Structural determinations from diamond drilling: Econ. Geology, v. 38, p. 298–312.

Rule, J. T., and Coons, S. A., 1961, Graphics: New York, McGraw-Hill Book Co., 484 p.

Stein, H. A., 1941, A trigonometric solution to the two-drill-hole problem: Econ. Geology, v. 36, p. 84–94.

Wisser, E., 1932, An aid in the interpretation of diamond drill cores: Econ. Geology, v. 27, p. 437–449.

Woolnough, W. G., and Benson, W. N., 1957, Graphical determination of the dip in deformed and cleaved sedimentary rocks: Jour. Geology, v. 65, p. 428–433.

CHAPTER 11

Stereographic Projection

A STEREOGRAPHIC PLOT is a simple way to study the angular relationships of planes and lines in space, such as rake, plunge, apparent dip, and the effects of rotation on the attitude of structures. It is impossible to represent problems involving distances on a stereographic projection. Consequently, problems of thickness of beds, or depth to them, or determinations of the amount of translation movement on a fault cannot be solved, because they involve computation of distances in addition to angles.

Properties of Stereographic Projection

The stereographic projection (Fig. 11–1) is one of the many projections used to represent a sphere on a plane surface. This is the same projection that is used to plot angular relations in crystallography, where it is referred to as a Wulff net. The identical device in structural geology is called a *stereonet*.

Figure 11–2 shows the principles involved in producing the stereonet of Figure 11–1. This net can be obtained from a sphere with meridians and parallels marked off in the conventional north-south and east-west positions. Imagine that a point source of light is located at 0° longitude and 0° latitude and that a plane is tangent to the sphere at 180° longitude and 0° latitude. The projection onto the plane of the hemispherical surface opposite the point source of light (hemisphere bounded by 90° east and 90° west meridians) produces the stereographic projection.

A stereographic projection can also be produced from a sphere by using a plane passing through the 90° east and west meridians and by a reference point located at 0° longitude and 0° latitude. The point of intersection of the plane with a line connecting the reference point with a specific point on the opposite hemisphere is the stereographic projection of the specific point. For structural geology purposes, we are interested in the hemisphere of possible orientations of dipping planes and plunging lines below a horizontal surface, so the plane of projection is in a horizontal attitude (Fig. 11–2). Meridians are north-south lines on the lower hemisphere. The curved lines trending generally east-west on the stereonet will be referred to as parallels. The meridians and parallels on the lower hemisphere project onto the horizontal plane as the stereonet of Figure 11–1.

Two very important characteristics of the stereographic projection are:

1. The projection of a circle on the sphere is also a circle on the stereographic projection.

2. Angles on the stereographic projection are equal to corresponding angles on the sphere. The trend and plunge of every line on the stereonet can be read in the same manner as spherical angles are read on a map.

FIG. 11–1 (below) A meridional stereographic projection or stereonet, drawn with 2° intervals. *(From D. J. Fisher with permission.)*

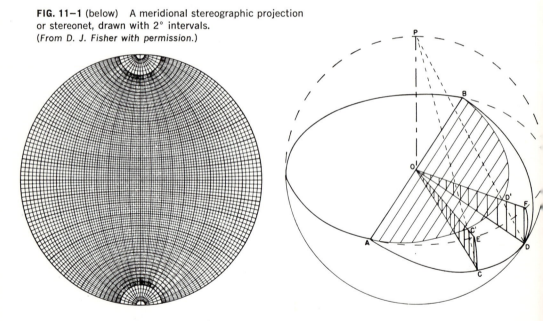

FIG. 11–2 (above right) Principle of stereographic projection and related functions. Point **P** is reference point for the projection. Points **A, E, F,** and **B** lie in the equatorial plane that bisects the sphere and is equivalent to the peripheral circle of Figure 11–1. Plane **ACDB** represents a dipping plane. Plane **AC′D′B** represents the stereographic projection of the dipping plane. Planes **FOD** and **EOC** are vertical. Line **AOB** represents the strike of the dipping bed. **OF** is true dip direction. Angle **FOD** is true dip amount. **OE** is apparent dip direction. Angle **EOC** is apparent dip amount. Measurements on the stereonet of Figures 11–1 and 11–9: **FD′** is plot of true dip amount; **EC′** is plot of apparent dip amount. If line **OC** is taken to represent the intersection of two planes; Angle **EOC** is plunge amount, (**EC′** is plot of plunge amount on stereonet); Angle **AOC** is rake in plane **ACDB** (**AC′** is plot of rake in plane **AC′D′B**, the stereographic plot).

Figure 11–3 is a stereographic projection like Figure 11–1, except that a 10° grid spacing is used, with dots at the intersection of every 10° increment of meridians and parallels. This coarser grid is used for illustrating stereonet procedures, because it yields uncluttered drawings.

FIG. 11–3 Ten-degree intersection point plot of stereonet in Figure 11–1, as used in this chapter for illustrations of problems.

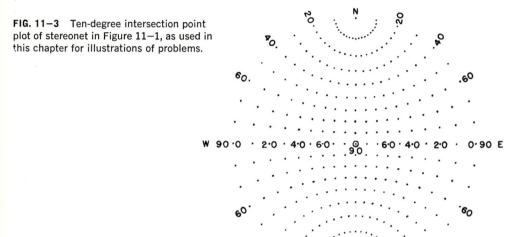

Figure 11–4 is an enlarged stereonet for use with the exercises at the end of this chapter. The large stereonet should be pasted on thin cardboard for permanent mounting. Then the sharp end of a thumb tack should be inserted from the back through the center of the stereonet, and a sheet of tracing paper centered and impaled on the tack should be used as an overlay on the stereonet. All lines plotted by use of the stereonet will be drawn on this overlay sheet. The compass direction of any line passing through the center of the stereonet is plotted by connecting the center of the projection with the point on the periphery of the projection corresponding to the direction of that line.

The first step in solving any problem by use of a stereonet is to draw a north-south line on the overlay, passing through the center of the projection. In the illustrations of this chapter, this reference line is marked by a short line labeled "N" at the periphery of the projection. Hereafter, the overlay sheet will be said to be in the *starting position* when the north-south line on the overlay sheet is coincident with the north-south line on the underlying stereonet. Directions of lines on the overlay sheet should be read only when the overlay sheet is in the starting position.

Because directions, but not distances, can be plotted on the stereonet, the attitude of any plane or line is plotted as if that plane or line passed through the center of the sphere. Figures 11–5 and 11–6 illustrate the procedure for plotting the attitude of a plane on the stereonet. The example used is a plane with an attitude of N48°W 27°S42°W. With the overlay sheet in the starting position, draw a line passing through the center of the projection representing the strike of the plane (N48°W). Draw a line

FIG. 11–5 Position for tracing a dip amount of 27° for bed having attitude N48°W 27°S42°W.

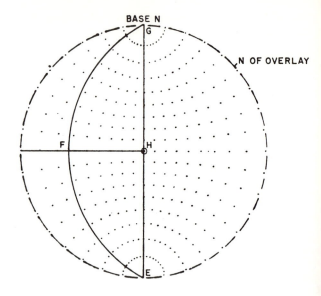

FIG. 11–6 Stereographic plot of plane with attitude N48° 27°S42°W.

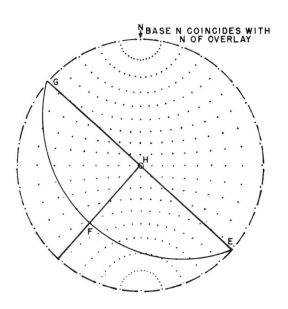

perpendicular to the strike representing the dip direction. Turn the strike line to a north-south position, and trace the meridian representing a dip of 27°. The position for this tracing is shown in Figure 11–5. Dip values increase from the periphery to the center of the stereonet. This traced meridian and the strike line represent the dipping plane (EFGH). The overlay sheet is then moved back to the starting position, and the plane is represented in its proper orientation in space as shown in Figure 11–6. All planes should be plotted in this manner.

FIG. 11–4 Large stereonet for solving problems at end of chapter.

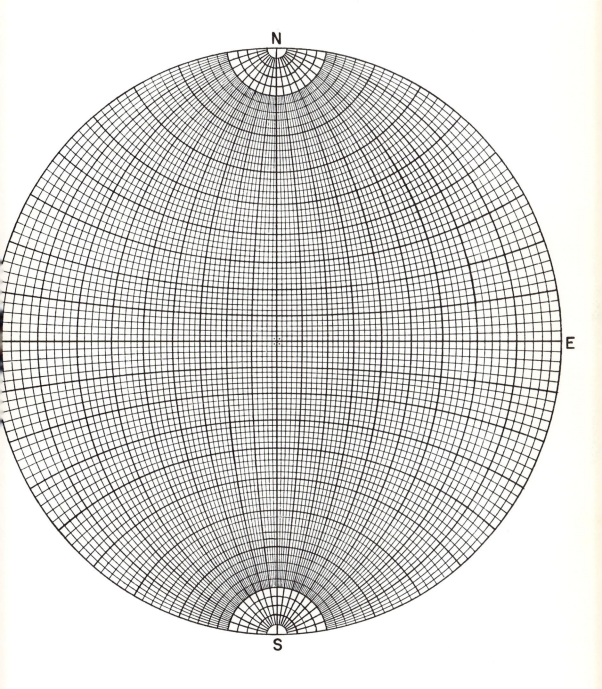

Figures 11–7 and 11–8 illustrate the procedure for plotting a line on a stereonet. Assume that we wish to plot a line plunging 22° in a direction S48°W. First, draw a faint line from the center of the stereonet to the position of S48°W on the periphery. Rotate the overlay sheet 42° clockwise until the radial line is an east-west position. Then count in 22° from the periphery along the east-west line, placing a point at the position where the plunge line pierces the sphere (Fig. 11–7). Rotate the overlay back to the starting position, and the piercing point is properly oriented for a line plunging 22°S48°W (see Fig. 11–8).

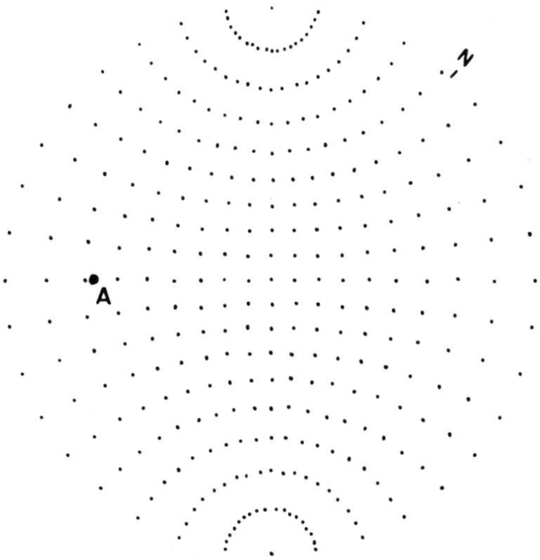

FIG. 11–7 Position for plotting plunge amount of line plunging 22°S48°W.

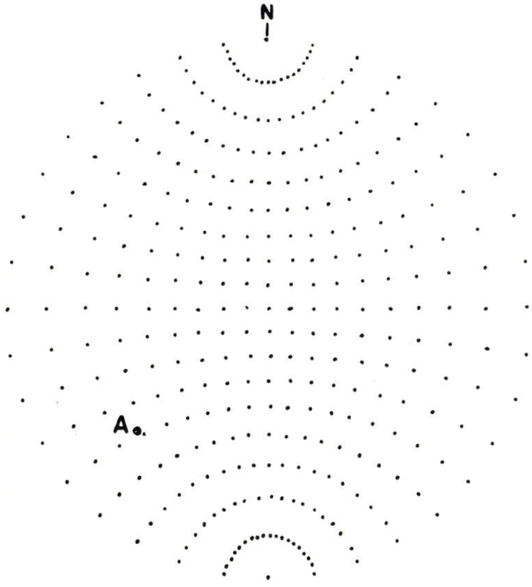

FIG. 11–8 Stereographic plot of line plunging 22°S48°W.

Many types of geological problems can be solved through stereonet analysis if these basic principles are remembered. The procedures described in the remainder of this chapter are among the most useful. For a more elaborate treatment of stereonets the reader is referred to the bibliography at the end of the chapter, particularly the general articles by Bucher (1944) and McClellan (1958) and the books by Badgley (1959), Haman (1961), and Phillips (1960). Higgs and Tunell (1959) and Phillips (1960) give some of the spherical trigonometry basis for stereonet graphical analysis. The special applications of most of the other references are indicated by their titles.

Apparent Dip from True Dip

Example problem—The true dip of a bed is 27°S42°W. What is the apparent dip in a direction N65°W? (Reference Fig. 11–9.)

Draw the north-south reference line on the overlay sheet. Plot the attitude of the bed by the method described above. On the overlay sheet draw a line extending N65°W from the center of the stereonet. Rotate the overlay sheet until the line originally drawn in a direction N65°W is in the due-west position (Fig. 11–9). At the intersection of the apparent dip direction line and the projected circle of the plane (point J), read the apparent dip angle of the bed in the direction HJ (8½°).

True Dip from Two Apparent Dips

Example problem—Two apparent dips of a bed are 36°S27°W and 23°N62°W. What is the true dip of the bed? (Reference Fig. 11–10.)

Draw the north-south reference line on the overlay sheet. Draw a line S27°W in the direction of the apparent dip. Rotate the overlay sheet until this apparent dip direction is in the east-west position. Place a mark (K) on the apparent dip direction line at its intersection with the meridian equal in value to the apparent dip. Do not complete the meridian, but instead just mark the apparent dip (plunge) by point K. Use the same procedure to plot the other apparent dip (point J in the direction HJ). Then rotate the overlay sheet until points K and J lie on the same meridian. Draw that meridian passing through K and J, and draw the strike line GHE. Plane GHEKFJ is the plane that contains both the true and apparent dips. Read the value of the meridian (GJFKE) containing both K and J, which is the true dip amount of 41° (Fig. 11–10). Rotate the overlay sheet until it is in the starting position, and read the true dip direction (the direction of line FH is S59°W). The true dip of the bed is thus 41°S59°W.

When solving a three-point problem (Fig. 5–2) with the stereonet, choose the two steepest of the three possible apparent dips for plotting on the stereonet. If this is done, a more precise answer can be obtained than with the gentlest of the three possible apparent dips.

Intersection of Two Planes

The intersection of any two planes is a straight line whose attitude can be described by its plunge or the rake of that line on the surface of one of the planes. A graphical solution for the line of intersection was developed in Chapter 8 (Fig. 8–2), and Formulae 58 and 60 in Chapter 9 can be used to compute the rake and plunge of the line. A stereonet solution is also available.

Example problem—Plane A dips 36°S27°W and plane B dips 23°N42°W. Determine the plunge of the line of intersection of the two planes. (Reference Fig. 11–11.) What is the rake in plane A of the line of intersection of the two planes? (Reference Fig. 11–12.)

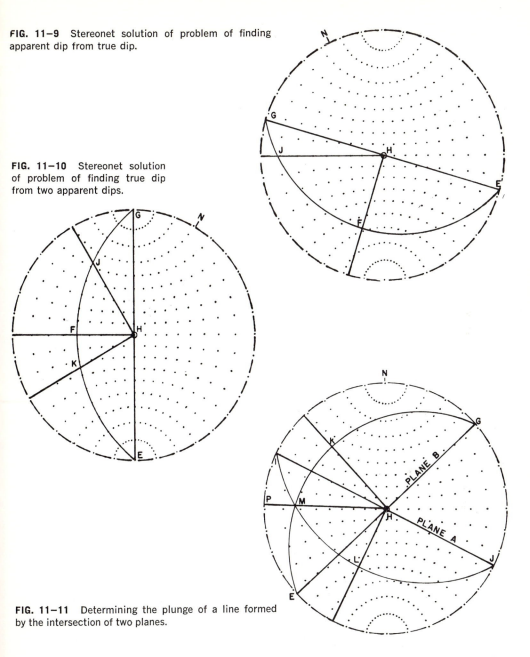

FIG. 11–9 Stereonet solution of problem of finding apparent dip from true dip.

FIG. 11–10 Stereonet solution of problem of finding true dip from two apparent dips.

FIG. 11–11 Determining the plunge of a line formed by the intersection of two planes.

Draw the north-south reference line on the overlay sheet (Fig. 11–11). Plot the attitude of plane A (IMLJH) and plane B (GKMEH). These two planes intersect at point M, and the direction of line HMP read on the periphery of the stereonet is the plunge direction of the line of intersection of the two planes (N88°W). In order to determine the plunge amount, rotate the overlay sheet until line HMP is in an east-west position on the stereonet, and read the value of the meridian on which point M lies (17°). The plunge of the line of intersection of the two planes is thus 17°N88°W.

In order to determine the rake in plane A of the line of intersection of the two planes, rotate the overlay sheet until the strike of plane A is in a north-south position (Fig. 11–12). Read the value of the parallel containing point M (34°). This is the rake of line HM in plane HIMLJ. The values for the parallels as shown in Figure 11–12 increase from 0° at the intersection of the north-south meridian with the periphery of the stereonet to 90° at the east-west parallel.

This technique can be used to determine the rake and plunge of the line of intersection of any two planes such as beds, dikes, veins, or faults. It is particularly useful for finding the rake in a fault plane of any of these which intersect a fault. Such measurements can in turn be used to find the relative movements of the fault, as shown in Chapter 9.

FIG. 11–12 Determining the rake of a line formed by the intersection of two planes.

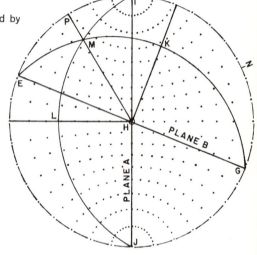

Tilting Problems

Rotation about Horizontal Axis / *Example problem*—The beds above an unconformity dip 30°S45°E. The beds below the unconformity dip 56°S8°E. What was the dip of the older beds when the beds above were being deposited, assuming that the upper beds accumulated as horizontal strata? (Reference Figs. 11–13 through 11–15.)

Draw the north-south reference line on the overlay sheet. Plot the attitude of the beds above the unconformity (GKEH in Fig. 11–13) and those below the unconformity (ILJH), using the conventional manner for plotting planes.

Rotate the overlay sheet until the strike (GHE) of the beds above the unconformity is in a north-south position (Fig. 11–14). The effect of tilt of the younger beds after their deposition can be removed by rotating plane GKEH until it is horizontal and GKE is in the position GTE. This is done by moving the loci of the intersections of the planes and the sphere each 30° meridian value to the east of the positions shown for GKEH and JLIH in Figure 11–14. GKE is thus moved to position GTE. The locus of points along ILJ is also moved 30° meridian value to the east (value of the parallels remaining the same) to the position shown by the new series of small crosses located 30° east of the corresponding crosses marking points on ILJ.

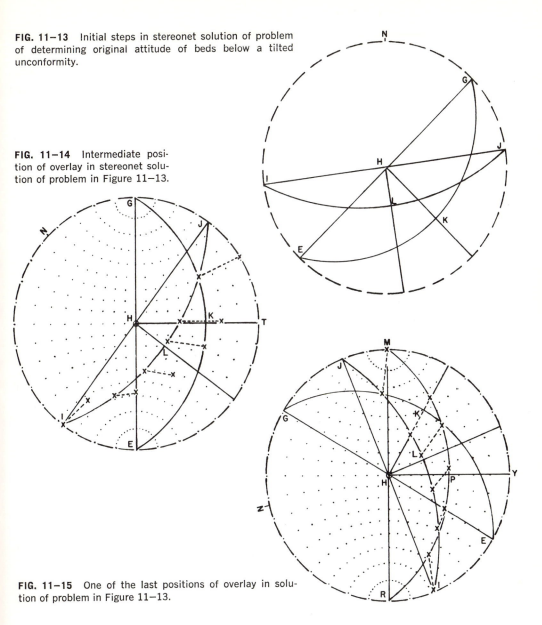

FIG. 11–13 Initial steps in stereonet solution of problem of determining original attitude of beds below a tilted unconformity.

FIG. 11–14 Intermediate position of overlay in stereonet solution of problem in Figure 11–13.

FIG. 11–15 One of the last positions of overlay in solution of problem in Figure 11–13.

This new series of small crosses (representing the position of IHJL before tilting of the beds above the unconformity) is rotated about the piercing thumb tack until all of the new small crosses are on the same meridian (Fig. 11–15). This meridian (MPR) is then traced through the small crosses. This is the plot of the attitude of the beds below the unconformity at the time of deposition of the beds above. The dip amount is read directly at point P (36°). In order to read the dip direction, rotate the overlay sheet to the starting position and read the direction of HPY on the periphery of the stereonet. The attitude of the beds below the unconformity at the time of deposition of the younger beds was thus 36°S14°W.

Another procedure may be applied to this two-tilt problem. One way to represent a plane in space is to plot the intersection trace of the plane with a stereonet sphere, as has been done in earlier parts of this chapter. A plane can also be uniquely oriented by plotting the location of the point where a line normal to that plane pierces the stereonet sphere. In the example problem just described, a line normal to the tilted unconformity surface plunges 60°N45°W, and a line normal to the bedding plunges 34°N8°W. These two lines are plotted in Figure 11–16. Rotate the overlay sheet until point A representing the bed above the unconformity is on an east-west line (Fig. 11–17). Rotate the position of A to the center of the stereonet at C (piercing point for a line normal to a horizontal plane). Rotate B along its parallel through an angle equal to the angle between A and C (30°). The new location of B (point E) corresponds to the attitude of the bed below the unconformity before tilting of the beds above the unconformity. Rotate the overlay back to its starting position, and determine the plunge direction and amount of the line represented by point E. From the plunge of a line normal to the original bedding before the unconformity was tilted, calculate the strike and dip of the strata beneath the unconformity at the time of deposition of the beds above the unconformity. The pre-unconformity attitude of the beds beneath the unconformity is 35°S15°W, which agrees well with the earlier solution obtained by a different method.

Fault Rotation about Inclined Axis / In addition to problems of rotation about a horizontal axis, stereonets can also be used to study rotation about an inclined axis. The most common situation involves simple rotational movement of an inclined fault. Of course, any possible translational movement cannot be ascertained from stereonet investigations.

Example problem—A fault has an attitude of N30°E 45°SE. The beds on the northwest side of the fault dip 35°N10°W. The beds on the southeast side of the fault dip 17°N40°W. Assuming that the beds on the northwest side have the same attitude as before faulting, what is the rotational movement of the fault? (See Fig. 11–18 for solution.)

Plot the piercing points of lines normal to the fault (point A, representing a line plunging 45°N60°W), the northwest set of beds (point B, corresponding to a line plunging 55°S10°E), and the southeast set of beds (point C, indicating a line plunging 73°S40°E). Move the whole stereographic projection of points in space so that the fault plane is horizontal. Rotate the overlay sheet so that point A lies on the east-west line (Fig. 11–18), and then move point A 45° along the east-west great circle to the vertical position (point A'). The fault plane is thus rotated to a horizontal position. Points B and C move 45° along small circles to B' and C'. Points B' and C' are a stereographic projection of the piercing points of lines normal to the beds when viewed in such a manner that the fault plane is perpendicular to the line of sight. The fault rotation was about a line normal to the fault plane, that is, about the line represented by A'. The angular rotation of the fault can be measured on the periphery of the stereonet. Transfer the latitude of B' and C' along their respective parallels to the periphery of the stereonet (points B" and C"). The angle between B" and C" is 20°, read on the peripheral circle. This is the amount of rotation on the fault which produced the observed attitudes of the beds. The hanging wall rotated 20° counterclockwise relative to the footwall.

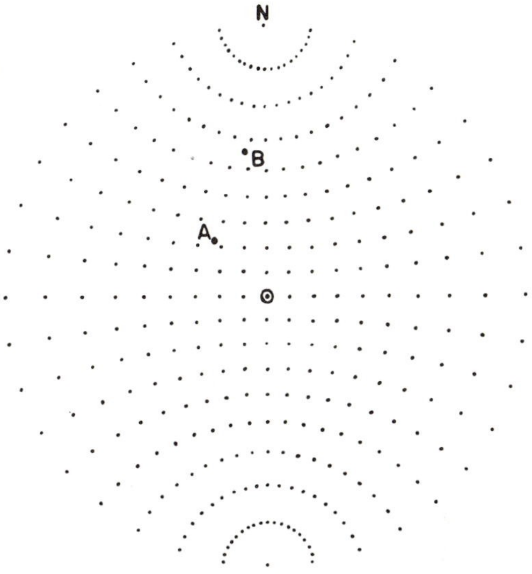

FIG. 11–16 Initial plotting in alternate method in two-tilt problems.

FIG. 11–17 Secondary overlay position in solution of problem in Figure 11–16.

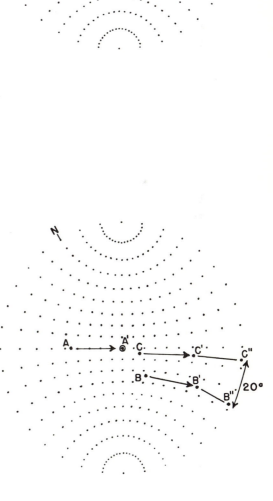

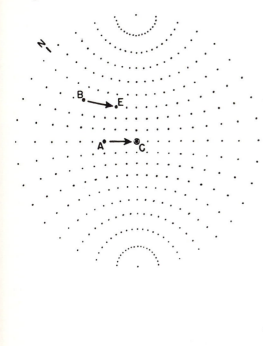

FIG. 11–18 Rotation along inclined fault plane.

Attitude of Beds from Core Data

In Chapter 10 descriptive geometry solutions were presented for determining possible attitudes of bedding, based on the angle bedding makes with the axis of inclined cores. These graphic procedures were quite complex, although they had the advantage of providing a pictorial result. The stereonet method is quicker, and probably as accurate as the descriptive geometry solution.

Example problem—Core 1 plunges 50°N45°E, and the bedding makes an angle of 55° with the core axis. Core 2 plunges 55°N65°W, and the bedding makes an angle of 35° with the core axis. Core 3 plunges 40° due south, and the angle between the bedding and core axis is 20°. What is the attitude of the bedding? (Solution in Figs. 11–19 and 11–20.)

The angles between the core axis and a line normal to the bedding plane are 35°, 55°, and 70° for cores 1, 2, and 3, respectively. For each core this angle rotated about the core axis describes a double cone (Fig. 10–4), and the line normal to the true bedding plane is an element common to each of the three double cones. One proceeds by finding the elements in common to the first and second cones, and then the elements in common to the second and third cones. The one element that is common to these two solutions is a line normal to the true attitude of the bedding. As a check, the first and third cones can be analyzed, also resulting in the correct answer obtained by the other two comparisons. Details of the method are as follows:

1. Plot on the overlay sheet (Fig. 11–19) the piercing points of the axes (points 1 and 2) of cores 1 and 2.
2. Rotate the stereonet until points 1 and 2 lie on a common meridian (66°), and draw that meridian on the overlay. This is the stereographic projection of the plane containing the two core axes. Then rotate the sphere 66° about the north-south axis so that 1′ and 2′ lie on the periphery of the stereonet. The plane containing the core axes is thus rotated to a horizontal position.
3. With the point 1′ in the north polar position, draw a small circle arc of 35° radius around the north and south poles. The 35° is the complement of the angle between the axis of core 1 and the bedding plane.
4. In like manner draw small circle arcs of radius 55° around point 2′ and its antipodal point.
5. The sets of small circle arcs intersect at points A and B. Rotate the overlay sheet until points 1 and 2 are again on a common meridian. Then rotate the stereonet 66° back to its original position. Point 1′ moves to point 1, 2′ to 2, A to A′, and B to B. Note that these movements take place along small circles. Point B moved completely off the stereonet on one side, and the point moved from the opposite edge to B′, a total movement of 66° along the small circle.
6. Points A′ and B′ are the stereographic projections of the elements common to the two cones described around cores 1 and 2. With the overlay returned to the starting position, one can determine that the two possible lines normal to true bedding are plunging 70°S76°E and 30°N4°E.

FIG. 11–19 Solution for inclined cores 1 and 2 in text example.

FIG. 11–20 Solution for inclined cores 2 and 3 in text example.

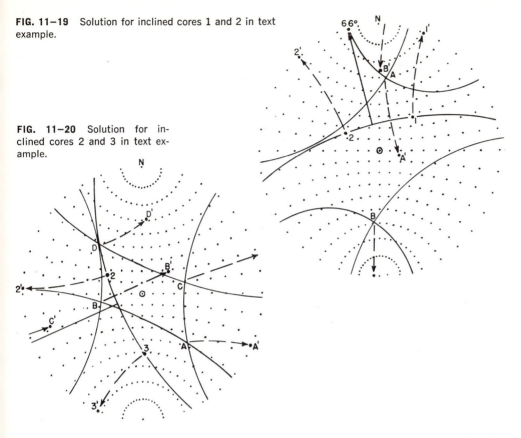

7. Follow the same procedure with cores 2 and 3 (Fig. 11–20). It happens that the small circle arcs intersect at four places, points A, B, C, and D. When these four points are rotated along with 2′ and 3′ so that 2′ and 3′ are at 2 and 3 again, then A′, B′, C′, and D′ yield four possible solutions for a line normal to the true bedding. These four solutions are 61°N53°E, 30°N5°E, 11°S72°W, and 1°S64°E.

8. An attitude of 30°N5°E for the line normal to the bedding plane is the only possible common solution for both cores 1 and 2 and cores 2 and 3. Therefore, 30°N5°E is the correct answer to the problem.

9. As a check, follow the same procedure for cores 1 and 3. Possible answers for the normal line are 30°N6°E, 69°N20°W, 36°N88°E, 18°N47°E. Note that approximately 30°N5°E is a solution for all three separate trials of two cores. Agreement within two or three degrees should be obtained, which is within the probable reliability of the initial data entering into the problem.

10. The attitude of a line normal to the true bedding is 30°N5°E. Therefore, the true bedding has a dip of 60°S6°W.

From the stereonet analysis of two cores, one may obtain one, two, three, or four possible solutions satisfying the data for those two cores. Use of a third core inclined differently permits determination of the one possible solution that gives the true attitude of the strata. If a common solution cannot be obtained from all three cores, then either a mistake in procedure or a fold (instead of a planar dip) is indicated. Use of the graphical solution given in Chapter 10 may help decide this alternative. Note that this method can be used for cores that would give rise to all three types of conic sections (ellipses, hyperbolas, and parabolas), and in many respects stereonets provide the easiest solution to problems of inclined cores.

Exercises /

249. The Madison Limestone dips 32°N78°W. What is its apparent dip in a direction S35°W?
250. The attitude of the Copper Ridge Dolomite is N37°E 48°SE. What is its apparent dip in a direction S5°W?
251. Two apparent dips in the Mauch Chunk Group are 13°N27°E and 25°N40°W. What is the true dip?
252. Two apparent dips are taken on an argillaceous sandstone. One is 15°S31°E, and the other is 27°S5°W. What is the true dip?
253. A bed dipping 35°S78°W intersects a vein dipping 57°N63°W. What is the plunge of the line of intersection of these two planes? What is the rake of the vein in the bed and the rake of the bed in the vein?
254. A dike has an attitude of N27°E 68°NW. An intersecting bed dips 23°N80°E. What is the plunge of the line of intersection of the two surfaces? What is the rake of the bed in the dike and the rake of the dike in the bed?
255. The attitude of a fault is N71°W 40°NE. An intersecting bed dips 29°S45°W. What is the plunge of the line of intersection of the two surfaces? What is the rake of the bed in the fault surface?
256. A vein dipping 69°N55°E intersects a fault dipping 29°S45°W. What is the plunge of the line of intersection of the two surfaces? What is the rake of the vein in the fault?
257. The beds below an angular unconformity dip 58°S52°E. A black, fissile shale above the unconformity dips 18°N80°W. What was the attitude of the beds below the unconformity when those above this surface were being deposited?
258. A geologist measures the dip of a fault as 55°S65°W. The fault movement pre-dated formation of an unconformity. The beds above this unconformity were deposited in a horizontal position but now dip 32°N35°E. What was the attitude of the fault at the time of deposition of the beds above the unconformity?
259. A joint set presently trends N81°E 72°NW in beds dipping 11°S10°E. The joints were apparently formed before warping resulted in the present attitude of the beds. What was the attitude of the joint set before warping?
260. A vertical fault strikes N75°E. Beds on the north side of the fault dip 35°N38°W. Beds on the south side of the fault dip 34° due north. Describe the rotation of the fault, assuming that the north block has not moved.
261. The dip of a fault plane is 30°N65°W. Beds on the northwest side of the fault dip 47°N10°W. The beds on the southeast side of the fault dip 10°N25°E. Assume that the beds on the northwest side remained fixed. How many degrees did the beds on the southeast side rotate?

262. A fault plane dips 56°S13°E. Beds on the north side of the fault dip 29°S70°W. Beds south of the fault are horizontal, and they presumably did not move during faulting. How many degrees did the north side rotate?

263. Solve exercise 245 on page 169 by stereonet.

264. Solve exercise 246 on page 169 by stereonet.

265. Solve exercise 247 on page 169 by stereonet.

266. Solve exercise 248 on page 170 by stereonet.

References

Badgley, P. C., 1959, Structural methods for the exploration geologist: New York, Harper and Bros., p. 187–242.

Beckwith, R. H., 1947, Fault problems in fault planes: Geol. Soc. America Bull., v. 58, p. 79–108.

Bucher, W. H., 1943, Dip and strike from three not parallel drill cores lacking key beds (stereographic method): Econ. Geology, v. 38, p. 648–657.

——, 1944, The stereographic projection, a handy tool for the practical geologist: Jour. Geology, v. 52, p. 191–212.

Clark, R. H., and McIntyre, D. B., 1951, A macroscopic method of fabric analysis: Am. Jour. Science, v. 249, p. 755–768.

Den Tex, E., 1953, Stereographic distinction of linear and planar structures from apparent lineations in random exposure planes: Jour. Geol. Soc. Australia, v. 1, p. 55–66.

DeWitte, A. J., 1956, A graphic method of dipmeter interpretation using the stereonet: Am. Inst. Min. Met. Eng., Tech. Paper 4333, Petroleum Trans. 207, p. 192–199.

Donn, W. L., and Shimer, J. A., 1958, Graphic methods in structural geology: New York, Appleton-Century-Crofts, Inc., 180 p.

Fisher, D. J., 1937, Some dip problems: Am. Assoc. Petroleum Geologists Bull., v. 21, p. 340–351.

——, 1938, Problem of two tilts and the stereographic projection: Am. Assoc. Petroleum Geologists Bull., v. 22, p. 1261–1271.

——, 1941, A new projection protractor: Jour. Geology, v. 49, p. 292–323, 419–442.

——, 1941, Drill hole problems in the stereographic projection: Econ. Geology, v. 36, p. 551–560.

——, 1943, Measuring linear structures on steep-dipping surfaces: Am. Mineralogist, v. 28, p. 204–208.

——, 1958, Graphical determination of the dip in deformed and cleaved sedimentary rocks: A discussion: Jour. Geology, v. 66, p. 100.

Gilluly, J., 1944, Dip and strike from three not parallel drill cores lacking key beds: Econ. Geology, v. 39, p. 359–363.

Haman, P. J., 1961, Manual of stereographic projection: Calgary, Canada, West Canadian Research Publications, 67 p.

Harrison, P. W., 1957, New Technique for three-dimensional fabric analysis of till and englacial debris containing particles from 3 to 40 mm. in size: Jour. Geology, v. 65, p. 98–105.

Higgs, D. V., and Tunell, G., 1959, Angular relations of lines and planes with applications to geologic problems: Dubuque, Iowa, William C. Brown Co., 48 p.

Hobson, G. D., 1943, A graphical solution of the problem of two tilts: Geol. Assoc. Proc., v. 54, p. 29–32.

——, 1945, The application of tilt to a fold: Geol. Assoc. Proc., v. 55, p. 216–221.

Ingerson, E., 1942, Apparatus for direct measurement of linear structures: Am. Mineralogist, v. 27, p. 721–725.

Johnson, C. H., 1939, New mathematical and stereographic net solutions to problems of two tilts—with applications to core orientation: Am. Assoc. Petroleum Geologists Bull., v. 23, p. 663–685.

Lowe, K. E., 1946, A graphic solution for certain problems of linear structure: Am. Mineralogist, v. 31, p. 425–434.

McClellan, H., 1958, Stereographic problems, p. 531–551 in J. D. Haun and L. W. LeRoy, editor, Subsurface geology in petroleum exploration: Golden, Col., Colorado School of Mines, 887 p.

Phillips, F. C., 1960, The use of stereographic projection in structural geology: London, Edward Arnold Ltd., 2nd ed., 86 p.

Stockwell, C. H., 1950, The use of plunge in the construction of cross-sections of folds: Geol. Assoc. Canada Proc., v. 3, p. 97–121.

Wallace, R. E., 1948, A stereographic calculator: Jour. Geology, v. 56, p. 488–490.

CHAPTER 12

Palinspastic Maps

IT IS OBVIOUS WHEN STUDYING OROGENIC BELTS that the outcrops seen today represent rocks formed someplace else and then moved by orogenic forces to their present positions. Little attention was given to this observation until recent years, but now it is evident that structural restorations are useful in making a variety of geologic interpretations. This aspect of geologic investigation is still in its infancy, and the present chapter is intended to stimulate thinking about the pre-deformation position of rocks as an aid in understanding their geologic history.

Maps that restore present-day geology to its geographic position at some specific time in the geologic past are called *palinspastic maps,* a term derived from the Greek meaning "stretched back" (Kay, 1937, p. 291). Such a map was first prepared by Kay as an estimate of the geographic position in Middle Ordovician time of the thrust sheets that were formed in New England by the Late Ordovician Taconic orogeny.

In 1945, Kay described palinspastic maps in considerable detail and formulated some general principles for their construction. In that article he presented generalized palinspastic base maps of North America for the time just before the Taconic, Appalachian, and Nevadian orogenies. He advocated the great value of these maps in aiding regional stratigraphic interpretation. More detailed palinspastic maps of parts of the Appalachians have been prepared separately and explained jointly by Dennison and Woodward (1963). Dennison (1961) used a palinspastic base for stratigraphic studies of the Devonian Onesquethaw Stage in West Virginia and adjacent states. Palinspastic consideration of the Devonian position of present-day outcrops greatly aided interpretation of stratigraphic relations. Kay and Crawford (1964) have made palinspastic comparisons of the Paleozoic miogeosynclinal and eugeosynclinal facies in the over-thrust belt of central Nevada. Bishop (1960, p. 131–132) has pointed out the value of palinspastic maps of individual oil fields as an aid in interpreting small-scale stratigraphic changes that appear on isopachous or lithofacies maps.

Low (1950, p. 956) applied the term palinspastic to a map showing the thickness of a sedimentary unit restored to its pre-erosional or pre-truncation dimensions. Kay (1954, p. 917) rightly claimed priority for the original designation and use of the name.

Palinspastic Maps of Sedimentary Terrain

The value of a palinspastic approach is most obvious for interpreting sedimentary terrain, probably because geologists have worked intermittently with this palinspastic problem since the 1930's. Any palinspastic map is a geologic interpretation of the amount of deformation of the rocks, but this deformation can be postulated within reasonable limits if certain assumptions are made. The alternative to a palinspastic consideration is plotting the stratigraphy on a road map or some other present-day base map. A poor palinspastic map is probably better than none at all. Even a tentative palinspastic base is likely to place stratigraphic data points nearer their depositional positions than a present-day base. Fortunately, refinements in structural understanding permit improvements in palinspastic base maps, so that successively better restorations can be made.

Some reference lines from the present-day world should appear on a palinspastic map. These most commonly are political boundaries (townships, counties, or states), lines of latitude and longitude, and the restored position of present-day outcrop belts. Any palinspastic map of a stratigraphic horizon is a map of a stratigraphic plane projected onto paper. We usually think of an outcrop trace as the intersection of a bedding surface with the present ground surface, but an outcrop trace can be just as correctly considered the locus of points on a geologically dated bedding surface that will happen to intersect the ground surface of the twentieth century. A palinspastic map restores this locus of points in the twentieth century to the position on the planar bedding surface at some specified ancient time. When preparing a palinspastic map, some reference position that remained undeformed is chosen as a stable area from which to restore the outcrops of the deformed belt back into a pre-deformation position. After the outcrop traces are palinspastically restored, then the political boundaries or meridians and parallels are distorted to the positions those present lines would have occupied at the stated time in the geologic past.

If the area has been subjected to multiple deformation, then a separate palinspastic base should be prepared for strata deposited prior to each deformation period. Successively more ancient maps would be constructed, with the effects of older deformations added to the maps for younger periods of deformation. For instance, Kay (1945) prepared palinspastic base maps for pre-Nevadian, pre-Appalachian, and pre-Taconic stratigraphy, plotting on a continental scale. Effects of epeirogeny involve dominantly vertical movement, so new maps are not necessary for each period of epeirogeny. The palinspastic position of several stratigraphic units can be plotted on a single map, provided they were all deformed simultaneously. An example would be a map of northern Wyoming showing the outcrop traces of the top of the Jurassic, Pennsylvanian, and Ordovician systems and the unconformity between the Precambrian and Cambrian rocks, since their present-day map positions are affected by a single period of orogeny, the Laramide at the end of the Cretaceous.

Stratigraphic Control Method of Preparation / Kay (1945, p. 439–441) considered methods of estimating the displacement of thrust sheets. He proposed that if the stratigraphy is known well enough in both the relatively overthrust and underthrust (or stable) sheets, then stratigraphic maps of the two fault blocks can be shifted to a position that would produce the simplest regional stratigraphic pattern (Fig. 12–1). The necessary amount of shift is an estimate of the restoration required for a palinspastic map. This does not give a unique amount of shift, because there could also have been a component of motion parallel to the regional trend of the isopachs. The restoration shown in Figure 12-1 is a minimum value for the distance of foreshortening.

Movement can be uniquely determined if two sets of intersecting stratigraphic lines are present on the map. Such a situation in Figure 12–1 would be the inclusion of north-south trending sand-shale ratio lines in addition to the northwest-southeast trending isopachs already shown. The map of the thrust sheet could be shifted so that each set of stratigraphic trend lines is respectively parallel on opposite sides of the restored fault and isopleth values are continuous as projected across the restored fault. Satisfaction of these conditions would yield a unique value for palinspastic restoration amount.

Use of this stratigraphic method to estimate foreshortening requires considerable detailed stratigraphic knowledge to prepare isopachous or lithofacies maps on both sides of the fault. In practice it is almost impossible to acquire sufficient information. In order to get the true directional trend of the isopach lines in Figure 12–1, it is essential to have outcrops scattered at different distances from the fault in each outcrop belt. Such a combination of distinct fold belts in each fault block rarely occurs in nature.

The method is impossible to use if only a single outcrop belt occurs on each side of the fault, because then it is impossible to determine the directional trend of the isopachs. Such a situation would be obtained in the faulted Appalachians of eastern Tennessee, where commonly only a single outcrop belt of southeastward dipping strata occurs in each fault block.

The minimum possible amount of foreshortening by a thrust is the distance between the farthest forward klippen or salient of the fault and the farthest back fenster or reentrant of the fault. This is true, unless the fault is simply a bedding-plane fault with the same strata always superimposed on opposite sides of the thrust. A bedding-plane fault with a few feet of displacement could have salients and recesses

FIG. 12—1 Method suggested by Kay (1945) to use stratigraphic evidence for estimating amount of horizontal displacement along a thrust fault. Dashed lines are traces of present-day outcrops. Dots are control points for preparing the isopachs, shown as numbered lines.

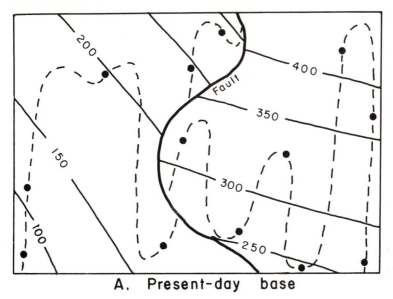

A. Present-day base

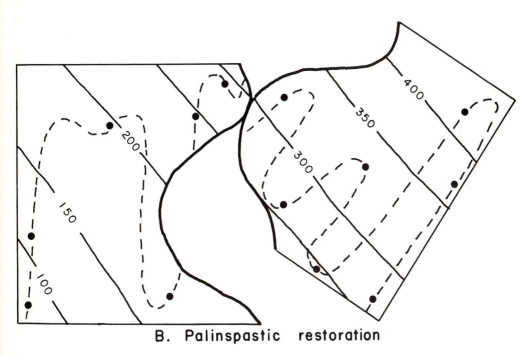

B. Palinspastic restoration

0   10
MILES

many miles in dimension if the beds and fault are nearly horizontal and then dissected by a dendritic drainage pattern.

The total displacement of any fault always equals or exceeds the stratigraphic displacement. Of course, a vertical fault would have no apparent horizontal displacement on a horizontal-plan palinspastic map, so the dip angle of the fault must be considered when preparing a palinspastic map.

Circular-arc Method of Preparation / Woodward postulated as early as 1936 that the amount of foreshortening in an orogenic belt is about 36% from the folding, assuming that the folds approximate semicircles (Dennison and Woodward, 1963, p. 672–673). His logic is illustrated in Figure 12–2. Arc CD is 1.57 times as long as straight line CD (½ π times diameter). The foreshortening by the folding is 36%. Similarly, the foreshortening of folds CDE is also 36%. The restored length of a quarter-circle is ½ π times the radius, still a foreshortening of 36%. In Figure 12–2 folds DEFG are composed of a series of quarter-circles, and the foreshortening is also 36% even though the fold is irregularly shaped. Many fold belts can be rather reliably approximated in cross section by a combination of quarter-circles, and Woodward's estimate would be reasonable if there had been no rock flowage.

Of course, Woodward also had to estimate the amount of horizontal displacement along faults and add that value to the distortion related to folding.

FIG. 12–2 Circular-arc method used by Woodward to estimate amount of foreshortening caused by folding.

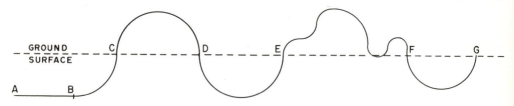

Sinuous-bed Method of Preparation / Figure 12–3 is a hypothetical map of an orogenic belt. Three structural sections are shown by lines X-X′, Y-Y′, and Z-Z′. Figure 12–4 is structural section Z-Z′. The before-folding position of A relative to B is obtained by measuring the length of curved line AB and plotting B at this straight-line distance to the right of A. Then the BC segment of the sinuous fold is measured, and C is plotted at BC straight-line distance to the right of B. This is continued completely across the map for all outcrop belts of the reference stratum. The horizontal component of fault HF is restored, and then HG is straightened to get the restored position of G relative to F. County boundary V is restored relative to C by using the same procedure, by locating the restored position of a spot on the reference horizon just beneath the ground surface at point V.

The position of the outcrop belt at D is restored on the section Z-Z′ (on both Figs. 12–3 and 12–4), and the comparable point on the outcrop belt is also restored on

FIG. 12–3 Hypothetical geologic map to illustrate procedures of palinspastic map construction with the aid of structural sections.

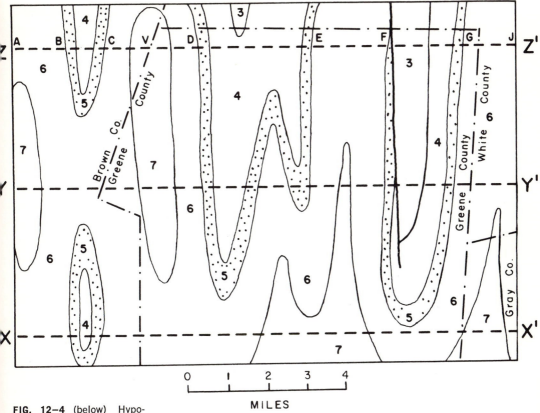

FIG. 12–4 (below) Hypothetical structural section Z-Z' located on the map of Figure 12–3.

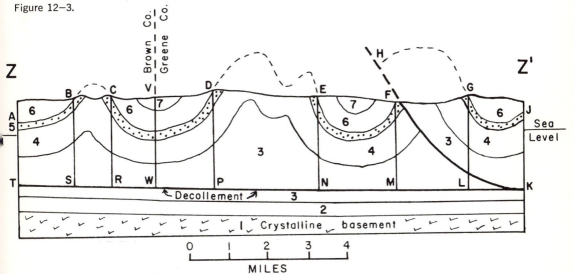

section Y-Y' making a sinuous-bed restoration along that structural section. Then the restored outcrop belt is interpolated on the palinspastic map between sections Z-Z' and Y-Y'. All outcrops of the reference stratum are palinspastically determined by using control points on structural sections and then interpolating outcrops between the lines of sections. If the rectangular shape of the map of Figure 12–3 is one bounded by meridians and parallels, the reconstructed shape of the rectangle is the palinspastic "restoration" of those grid lines.

The sinuous-bed method assumes that there has been no distortion of distances by flowage of strata during folding. Flowage is absent only in concentric folding. The folded Appalachians are sometimes considered as characteristic concentric folds, but even there the assymetry of the folds implies considerable rock flowage, even though it has not been intensive enough to produce cleavage in many places. The sinuous-bed procedure is probably the best method for palinspastic map construction if insufficient information is available for using the equal-area method.

Writers who have advocated the sinuous-bed method for estimating the amount of foreshortening of folds include Van Hise (1898, p. 10–12), Chamberlin (1910, p. 233), Cloos (1940, p. 849), Kay (1945, p. 438), and Scheidegger (1963, p. 20). Van Hise (1898, p. 10–41) and Chamberlin (1919, pp. 231–232) have considered sources of error in the sinuous-bed method and describe procedures for obtaining the most reliable estimates of foreshortening. These include:

1. So far as practicable, the same formation should be measured throughout the section.
2. The strongest (most competent) formations available should be selected for measurement.
3. Assume the least possible deformation in the subsurface where the reference horizon is not exposed.
4. Make the gentlest possible estimate of the folding of the reference horizon in areas where it has been removed by erosion.
5. Make estimates of foreshortening on the stratigraphic horizon for which the most structural control is available. Criteria indicating good control include abundant outcrops, frequent well penetration, and resistance to erosion so that accurate strikes and dips are abundant.
6. Measure at some distinct formation boundary instead of along a gradational formation contact.

Equal-area Method of Preparation / Bucher (1955, p. 357) expressed the opinion that in the Alps there had been such severe rock flowage that a sinuous-bed approach to amount of foreshortening was erroneous. He suggested that all significant deformation had been at right angles to the trend of the orogenic belt, with no appreciable deformation along the trend. He assumed that the rocks in the deformed portion of the crust maintained constant volume and cross-section area even though they were orogenically crumpled by flowage.

Dennison (1961, and in Dennison and Woodward, 1963) developed Bucher's concept into a specific method for estimating foreshortening of a structural section (see

Fig. 12–4). If a particular stratigraphic horizon is considered to localize a decollement (such as a stratigraphic position in Figure 12–4 about one-third of the distance above the base of formation 3), the elevation of the locus of the decollement is determined by the stratigraphic thickness between the sheared decollement stratum in a syncline and the known position of some reference horizon in the overlying part of the syncline. In Figure 12–4 the position of point W on the decollement is determined by the known position in that syncline of the superface of formation 5 in the structural section and the stratigraphic thickness of formations 4 and 5 plus the portion of formation 3 considered to lie above the decollement. The position of the decollement beneath each other major syncline is calculated in a similar manner, and the decollement surface TSRWPNMLK is drawn on the structural section. The palinspastically restored distance between point B relative to point A is calculated by measuring area ABST (using a planimeter or counting squares on translucent graph paper superimposed on the structural section), and then dividing that area by the stratigraphic thickness between the decollement and the top of formation 5, using stratigraphic thickness values in the major synclines where it is assumed that none of the formations was thickened by flowing. On a regional scale, synclines appear to be passive features, and the orogenic deformation raises up anticlinal structures or perhaps produces faults. Small errors in estimating the position of the decollement do not seriously affect the outcome of the palinspastic computation.

Restore point B relative to point A using the equal-area method. Then restore distance BC relative to C by using area SRCB, after projecting the beds back over the anticlinal structure to be included in the area measured. The same procedure is used to restore CD, DE, and EF. Restoring the faulted segment of G relative to F involves measuring the area bounded by fault FH, fold HG, and lines GL, LM, and MF. Then divide this area by the regional stratigraphic thickness between the decollement and the top of formation 5.

The palinspastic restoration of the county boundary at V relative to C involves measuring the area bounded by line RC, the top of formation 5 to the synclinal axis, the vertical line down to point W, and line WR, followed by the necessary division by stratigraphic thickness to yield restored distance between C and V.

Each structural section is treated this way to obtain restored position of outcrops of the reference bed along the section, and then restored outcrop traces are interpolated between the lines of structural section control. Political boundaries and latitude and longitude are also interpolated over the entire palinspastic map area.

Figure 12–5 shows structural control for a regional palinspastic map of a portion of the Ridge and Valley Province of the Appalachians. Figure 12–6 is a palinspastic restoration for the same area of outcrops of selected stratigraphic horizons, of state boundaries, and lines of latitude and longitude. The distortion produced by the Appalachian (or Allegheny) orogeny is most evident by comparing the shapes of state boundaries on the two maps and by noting the irregular shapes of meridians when plotted palinspastically. Figure 12–7 shows the amount of orogenic foreshortening in a simplified fashion. Additional interpretation of this example is given in the paper by Dennison and Woodward (1963).

FIG. 12–5 Published structural sections used to prepare palinspastic map of central Appalachians. (Figures 12–5, 12–6, and 12–7 are from "Palinspastic Maps of Central Appalachians" by the author and H. P. Woodward in *A.A.P.G. Bulletin,* April 1963.)

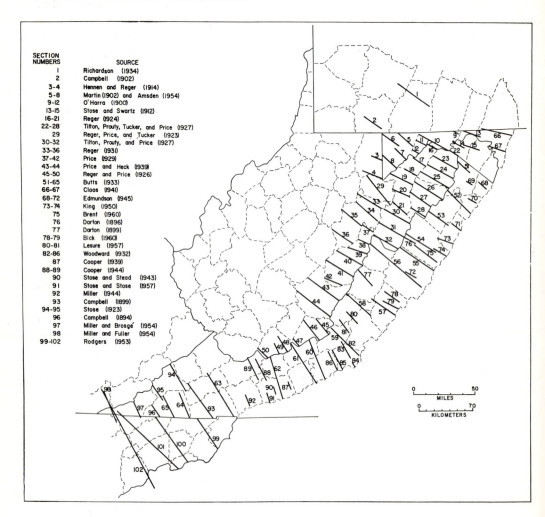

SECTION NUMBERS	SOURCE
I	Richardson (1934)
2	Campbell (1902)
3-4	Hennen and Reger (1914)
5-8	Martin (1902) and Amsden (1954)
9-12	O'Harra (1900)
13-15	Stose and Swartz (1912)
16-21	Reger (1924)
22-28	Tilton, Prouty, Tucker, and Price (1927)
29	Reger, Price, and Tucker (1923)
30-32	Tilton, Prouty, and Price (1927)
33-36	Reger (1931)
37-42	Price (1929)
43-44	Price and Heck (1939)
45-50	Reger and Price (1926)
51-65	Butts (1933)
66-67	Cloos (1941)
68-72	Edmundson (1945)
73-74	King (1950)
75	Brent (1960)
76	Darton (1896)
77	Darton (1899)
78-79	Bick (1960)
80-81	Lesure (1957)
82-86	Woodward (1932)
87	Cooper (1939)
88-89	Cooper (1944)
90	Stose and Stead (1943)
91	Stose and Stose (1957)
92	Miller (1944)
93	Campbell (1899)
94-95	Stose (1923)
96	Campbell (1894)
97	Miller and Brosgé (1954)
98	Miller and Fuller (1954)
99-102	Rodgers (1953)

Uses / The obvious use of a palinspastic map of sedimentary terrain is to restore the position of measured section localities and well penetrations back to their sites at the time of deposition of the stratum being studied. This is useful in interpreting sedimentology and geologic history. Nearby outcrop belts with dissimilar facies may have been deposited many miles apart. Rates of thinning of formations generally are less when considered on a palinspastic base. Similarly, the rate of facies change, such as

FIG. 12–6 Palinspastic map of central Appalachians, showing position of present outcrop belts of selected strata prior to deformation by Appalachian orogeny.

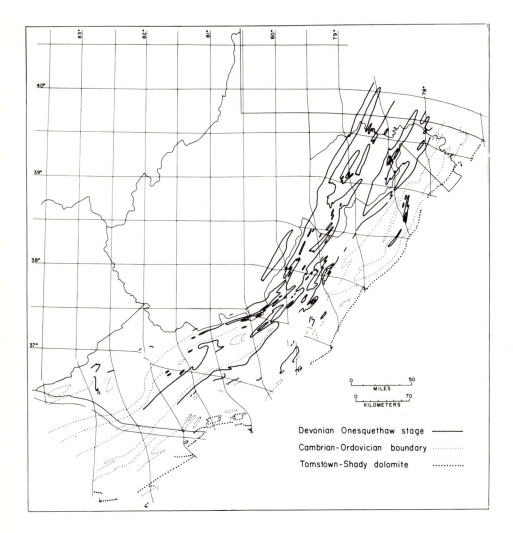

Devonian Onesquethaw stage ————

Cambrian-Ordovician boundary ··········

Tomstown-Shady dolomite ··········

sand-shale ratio, will usually be more gradual on a palinspastic base. The directional trend of an unconformity pinch-out of a formation may be modified on a palinspastic base. If there has been regional rotation of outcrop belts a few degrees about a vertical axis, current features such as cross-beds may not give a true mean current direction unless correction is made for the regional rotation of the strike of the beds in an outcrop belt.

FIG. 12–7 Orogenic foreshortening of Ridge and Valley Province expressed in miles of compression and per cent of shortening. Arrows delimit regions with severely deformed strata.

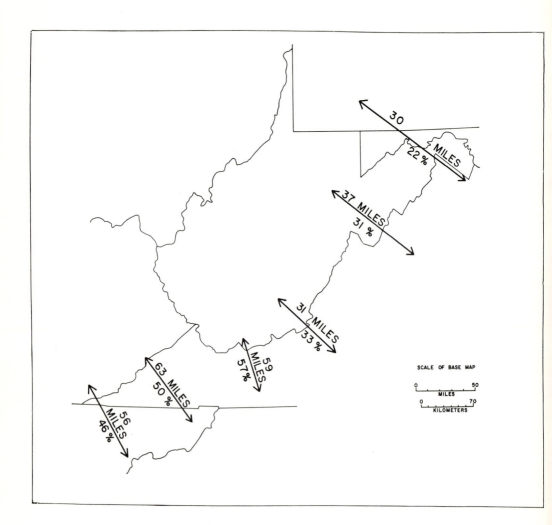

Palinspastic Maps of Igneous Terrains

Palinspastic maps of igneous terrain can be prepared to restore the shape of an intrusion to its original configuration before fragmentation by post-intrusion faulting. Perhaps the best example (Emeleus, 1964) is a map of present-day outcrops of a Precambrian igneous complex in Greenland with a palinspastic restoration of the shape of that intrusion before it was affected by a sequence of nearly vertical, strike-slip

FIG. 12–8 Structural development of the Grønnedal-Ika igneous complex, Greenland. Sketch 5 is present configuration of complex (shown by white area). Sketch 1 is calculated original shape of the intrusive complex, after effects of four faulting episodes are removed. (*C. H. Emeleus.*) (From C. H. Emeleus in Meddr. Grønland, Bd. 172, Nr. 3.)

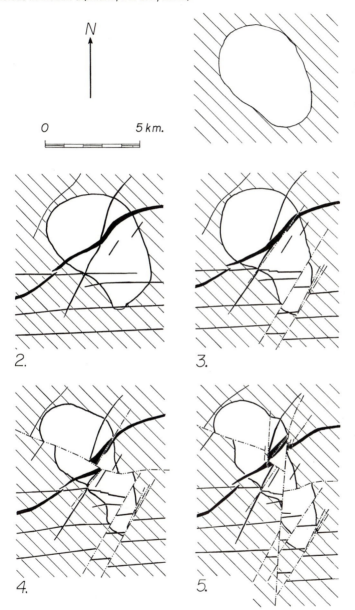

faults with dikes intruded along some of the fault surfaces. Both the present-day and palinspastic maps are beautifully colored in their publication. Figure 12–8 is a simplified diagram showing the palinspastic removal of successively more ancient faults that cut the igneous complex.

Exercises /

267. Prepare a palinspastic map of the Hollidaysburg-Huntingdon, Pennsylvania, folio (No. 227), using both the sinuous-bed and the equal-area methods. First draw a 1:250,000 scale reduction of the present-day geologic map for comparison with the palinspastic maps. On the present-day map show meridians and parallels at 5-minute intervals, county boundaries, and the outcrop trace of the Clinton Formation. Using the structural sections for control, prepare a 1:250,000 scale palinspastic map of the same reference lines for the time of Clinton deposition, following the sinuous-bed procedure. Start with a northwest-southeast line and then work toward the northeast and southwest corners. Prepare a comparable palinspastic map using the equal-area method, assuming a decollement 800 stratigraphic feet beneath the top of the Waynesboro Formation. Calculate the elevation of the decollement surface under the syncline at the northwest corner of the Hollidaysburg quadrangle and under the Trough Creek syncline in the Huntingdon quadrangle. Interpret the decollement surface as planar between the two reference points, extending it at a constant elevation along the direction of present-day regional strike. Compare the present-day base with the two palinspastic maps. Compare the two palinspastic maps with each other.

268. Prepare a palinspastic map of the Bessemer-Vandiver, Alabama, folio (No. 221), using both the sinuous-bed and the equal-area method. First draw a 1:250,000 scale reduction of the present-day geologic map for comparison with the palinspastic maps. On the present-day map show the meridians and parallels at 5-minute intervals, county boundaries, and the outcrop trace of the Chattanooga Shale, or its expected position where missing by unconformity. Using the structural sections for control, prepare a 1:250,000 scale palinspastic map of the same reference lines, for the time of Chattanooga Shale deposition, following the sinuous-bed procedure. Start with a northwest-southeast line of sections, and then work toward the northeast and southwest corners. Prepare a comparable palinspastic map using the equal-area method, assuming a decollement at the base of the exposed part of the Rome Formation. Obtain stratigraphic thicknesses from the columnar section on page 23 of the folio for figuring the elevation of the decollement surface, and assume that the level of the decollement is constant along present regional strike. Compare the present-day base with the two palinspastic maps. Compare the two palinspastic maps with each other.

References

Bishop, M. S., 1960, Subsurface mapping: New York, John Wiley & Sons, 198 p.

Bucher, W. H., 1955, Deformation in orogenic belts: Geol. Soc. America, Special Paper 62, p. 343–368.

Butts, C., 1927, Bessemer-Vandiver, Alabama Folio: U.S. Geol. Survey, Geol. Atlas of the U.S., folio 221, 22 p.

———, 1945, Hollidaysburg-Huntingdon, Pennsylvania Folio: U.S. Geol. Survey, Geol. Atlas of the U.S., folio 227, 20 p.

Chamberlin, R. T., 1910, The Appalachian Folds of Central Pennsylvania: Jour. Geology, v. 18, p. 228–251.

——, 1919, The building of the Colorado Rockies: Jour. Geology, v. 27, p. 145–164, 225–251.

Cloos, E., 1940, Crustal shortening and axial divergence in the Appalachians of southern Pennsylvania: Geol. Soc. America Bull., v. 51, p. 845–872.

Dennison, J. M., 1961, Stratigraphy of Onesquethaw Stage of Devonian in West Virginia and bordering states: W. Va. Geol. Survey, Bull. 22, 87 p.

——, and Woodward, H. P., 1963, Palinspastic maps of central Appalachians: Am. Assoc. Petroleum Geologists Bull., v. 47, p. 666–680.

Eardley, A. J., 1939, Slotted template for resolving crustal movements: Jour. Geology, v. 47, p. 546–554.

Emeleus, C. H., 1964, The Grønnedal-íka alkaline complex, South Greenland: Meddelelser om Grønland, Bd. 172, Nr. 3, 75 p.

Kay, G. M., 1937, Stratigraphy of the Trenton Group: Geol. Soc. America Bull., v. 48, p. 233–302.

——, 1945, Paleogeographic and palinspastic maps: Am. Assoc. Petroleum Geologists Bull., v. 29, p. 426–450.

——, 1954, Isolith, isopach, and palinspastic maps: Am. Assoc. Petroleum Geologists Bull., v. 38, p. 916–917.

——, and Crawford, J. P., 1964, Paleozoic facies from the miogeosynclinal to the eugeosynclinal belt in thrust slices, central Nevada: Geol. Soc. America Bull., v. 75, p. 425–454.

Low, J. W., 1950, Subsurface geologic maps and illustrations, p. 894–969 in L. W. Leroy, editor, Subsurface geologic methods: Golden, Colorado, Colorado School of Mines, 1156 p.

Scheidegger, A. E., 1963, Principles of geodynamics: Berlin, Springer-Verlag, 2nd ed., 362 p.

Van Hise, C. R., 1898, Estimates and causes of crustal shortening: Jour. Geology, v. 6, p. 10–64.

General References

Badgley, P. C., 1959, Structural methods for the exploration geologist: New York, Harper and Bros., 280 p.

———, 1965, Structural and tectonic principles: New York, Harper and Row, 521 p.

Beloussov, V. V., 1962, Basic problems in geotectonics: New York, McGraw-Hill Book Co., 809 p. (Translated from the Russian.)

Billings, M. P., 1954, Structural geology: Englewood Cliffs, N.J., Prentice-Hall, Inc., 2nd ed., 514 p.

Bucher, W. H., 1933, The deformation of the earth's crust: Princeton, N.J., Princeton University Press, 518 p.

Deetz, C. H., 1943, Cartography, a review and guide: U.S. Dept. of Commerce, Coast and Geodetic Survey, Spec. Pub. No. 205, 2nd ed., 84 p.

deSitter, L. U., 1964, Structural geology: New York, McGraw-Hill Book Co., 2nd ed., 551 p.

Donn, W. L., and Shimer, J. A., 1958, Graphic methods in structural geology: New York, Appleton-Century-Crofts, Inc., 180 p.

Eardley, A. J., 1962, Structural geology of North America: New York, Harper and Row, 2nd ed., 743 p.

Geike, J., Campbell, R., and Craig, R. M., 1953, Structural and field geology: London, Oliver and Boyd, 6th ed., 397 p.

Goguel, J., 1962, Tectonics: San Francisco, W. H. Freeman Co., 384 p. (Translated from the French.)

Higgs, D. V., and Tunell, G., 1959, Angular relations of lines and planes with applications to geologic problems: Dubuque, Iowa, William C. Brown Co., 48 p.

Hills, E. S., 1963, Elements of structural geology: New York, John Wiley & Sons, 483 p.

Hoelscher, R. P., and Springer, C. H., 1956, Engineering drawing and geometry: New York, John Wiley & Sons, 498 p.

Lahee, F. H., 1961, Field geology: New York, McGraw-Hill Book Co., 6th ed., 926 p.

Leith, C. K., 1923, Structural geology: New York, Henry Holt and Co., 390 p.

Leroy, L. W., and Low, J. W., 1954, Graphic problems in petroleum geology: New York, Harper and Bros., 238 p.

Lobeck, A. K., 1958, Block diagrams and other graphic methods used in geology and geography: Amherst, Mass., Emerson-Trussell Book Co., 2nd ed., 212 p.

Ludlum, J. C., 1956, Philosophical and geometric perspectives as aids in the teaching of structural geology: Jour. Geological Education, v. 4, p. 33–36.
——, and Dennison, J. M., 1960, Structural geology laboratory manual: Ann Arbor, Mich., Edwards Bros., 2nd ed., 145 p.
McIntyre, D. B., and Weiss, L. E., 1956, Construction of block diagrams to scale in orthographic projection: Geologists' Assoc. Proc., v. 67, p. 142–155.
McKinstry, H. E., 1948, Mining geology: Englewood Cliffs, N.J., Prentice-Hall, Inc., 680 p.
Metz, K., 1957, Lehrbuch der Tektonischen Geologie: Stuttgart, Germany, Ferdinand Enke Verlag, 294 p.
Moore, C. A., 1963, Handbook of subsurface geology: New York, Harper and Row, 235 p.
Nevin, C. M., 1949, Principles of structural geology: New York, John Wiley & Sons, 4th ed., 410 p.
Platt, J. I., and Challinor, J., 1954, Simple geological structures: London, Thomas Murby, 56 p.
Raisz, E., 1962, Principles of cartography: New York, McGraw-Hill Book Co., 315 p.
Roberts, A., 1958, Geological structures and maps: London, Cleaver-Hume Press Ltd., 2nd ed., 91 p.
Robinson, A. H., 1960, Elements of cartography: New York, John Wiley & Sons, 2nd ed., 343 p.
Rule, J. T., and Coons, S. A., 1961, Graphics: New York, McGraw-Hill Book Co., 484 p.
Russell, W. L., 1955, Structural geology for petroleum geologists: New York, McGraw-Hill Book Co., 427 p.
Scheidegger, A. E., 1963, Principles of geodynamics: Berlin, Springer-Verlag, 2nd ed., 362 p.
Slaby, S. M., 1956, Engineering descriptive geometry: New York, Barnes and Noble, Inc., 353 p. (College Outline Series.)
Stoves, B., and White, C. H., 1935, Structural geology with special reference to economic deposits: New York, Macmillan and Co., 460 p.
Warner, F. M., 1938, Applied descriptive geometry: New York, McGraw-Hill Book Co., 228 p.
Wellman, B. L., 1957, Technical descriptive geometry: New York, McGraw-Hill Book Co., 628 p.
Whitten, E. T. H., 1966, Structural geology of folded rocks: Chicago, Rand McNally and Co., 663 p.
Willis, B., 1923, Geologic structures: New York, McGraw-Hill Book Co., 2nd ed., 518 p.

Index